What's so special about *Special Relativity?*

How Galileo invented relativity and how Einstein improved it

W. David McComb

Paradox Popular Science

Contents

Preface

This is not a conventional textbook. In it you will not find exercises, diagrams or mathematical equations, with the pardonable exception of the most famous equation of all. My aim is to tell the story of what Einstein did, in as simple and as clear a way as possible. And to do so without mathematics. Instead, where necessary, we will use equations stated in words.

You will find out that in essence all Einstein did was to make one assumption that was new. But this was an assumption that flew in the face of everything that was understood about physics at the beginning of the twentieth century. Indeed, at the time, it was regarded not only as obviously wrong, but also as quite shocking. Yet, by forcing our description of the natural world to accommodate this one assumption, Einstein led us into a universe of marvels ranging from the initial creation of matter to its destruction in black holes.

The book is divided roughly into three sections. In the first section, in Chapter One, we briefly explain the basic difference between Galilean Relativity and Einstein's

Special Relativity. Then, the following Chapters, Two to Eight, set the scene, where we concentrate on giving a simple explanation of the Galilean-Newtonian picture of the universe, ending up with a brief account of Newton's laws of motion.

In the middle section of the book, we deal with the transition from classical physics to Einstein's relativity. It is now generally recognised that there were two routes leading from Galileo's relativity to Einstein's relativity.

First there were the inconsistencies which arose when the subject of electrodynamics (and, in particular, Maxwell's equations) was developed in the late nineteenth century. The entire essence of physics lies in its attempt to provide a unified and consistent description of nature. So the existence of inconsistencies suggests that something is wrong. We will discuss this aspect later on, in Chapter Nine.

Secondly, there was the null result of the Michelson-Morley experiment. In 1887, Michelson and Morley made two measurements of the speed of light: one in the direction of the Earth's motion, the other at right angles to it. In terms of the physics of the time, they expected that these two measurements would be different. Strictly speaking, they measured the difference between the two. And, as the two measurements were the same, their difference was zero. This is often referred to as 'the most important null result in the history of physics'. It is discussed in Chapter Ten.

Einstein took the first of these two routes and claimed to have been unaware of the second. In Chapters Eleven to Thirteen we discuss how Einstein's Special Relativity came about, including the introduction of the four-dimensional space-time continuum. In Chapter Fourteen we give a brief introduction to the ideas of General Relativity.

In the final section of the book, comprising Chapters Fifteen to Eighteen, we present an overview of the effects that Einstein's theories have had on the rest of physics. We begin with time dilation and the notorious 'twins paradox'. Then we go on to the mass-energy equivalence which explains how nuclei stick together, despite the strong electrostatic repulsion of the nucleons, and how stars use nuclear fusion to generate energy. This is followed by an account of its effects on our understanding of the universe, including the idea of 'black holes', and concludes with a look at the formalism which, when relativity is joined with the quantum extension of the formalisms of classical mechanics, underpins the whole of modern physics.

Lastly, for some chapters, additional material has been presented in chapter endnotes, including in some cases suggestions for further reading. These additional notes are collected together at the end of the book.

Chapter 1

What is relativity?

Many years ago I read an anecdote about Albert Einstein. He was staying in a rather grand hotel and, while waiting for the lift, was approached by a woman who gushed 'Oh, Professor Einstein, do tell me. What is Relativity?'. Einstein, who was about to step into the lift, did not reply. But, as the lattice-work doors were drawn across and the lift began to ascend, he looked down at the disappearing woman and said: 'Madam, this is relativity!'.

This reply, although strictly correct, may seem rather snubbing. Of course what the woman presumably meant was: 'What is this theory of yours that everyone is so excited about?'. Because she too, like many others, would have liked to join in the excitement. Quite understandable really, but also a trifle unrealistic to expect the great man to explain it all in a few words there and then.

This story was in a popular magazine, which is not necessarily the most unimpeachable of sources. Nevertheless, it illustrates an interesting point. The incident described is supposed to have taken place in the late 1940s, when Einstein was at the height of his fame. It has long been acknowledged that Einstein's popular renown exceeded that of any other scientist before or since. What this story illustrates is the extent to which his name was associated with the concept of relativity. For this reason it may come as a surprise to learn (and it certainly did to me) that the concept of relativity was not actually due to Einstein at all. It was in fact invented by Galileo, back in the sixteenth century. Indeed, there is a subject called Galilean Relativity which is of considerable importance in classical physics and its applications: for further information, see the notes for this chapter at the end of the book.

As we shall see, Galilean Relativity is equivalent to Newton's picture of the universe. Going beyond this, Einstein produced not one but two theories of relativity. These were, respectively, Special Relativity in 1905 and General Relativity in 1917. It was the second of these which led to his great fame. It made a prediction about the bending of light by the Sun, which could not be tested until the end of the Great War, when an expedition to the Island of Principe (off the coast of West Africa) led by Sir Arthur Eddington in 1919 measured the deviation of light from distant stars during a solar eclipse. The results were in good agreement with Einstein's prediction. This was seen as a spectacular achievement, and was reported on the front pages of most newspapers around the world. For the moment we may note that, loosely speaking, Special Relativity can be seen as a correction to Newton's laws of motion, whereas General Relativity can be seen as a correction to Newton's law of gravity.

We shall return to these considerations later. Now let us concentrate on our first, simplified look at relativity. At its very simplest, Galileo's recognition was that speed was relative. In other words, the speed of anything must be measured relative to something else. To understand this, it is helpful to compare the concept of speed with that of acceleration. We are all familiar with the feeling of acceleration. When the car we are in accelerates, we experience a force. This is in accordance with Newton's second law of motion (or N2, for short). However, the fact that we are in motion is less obvious to us when we are moving at constant speed. Of course we usually have plenty of visual clues. But in the absence of these, could we tell? In fact we will devote a considerable portion of the present book to persuading you that the answer to this question is 'no'.

Let us begin with a 'thought experiment' in which we have all the relevant information. Einstein originally defined this as an experiment which violates no known physical principle, but which cannot be carried out in the present state of technology. Such experiments are, in effect, exercises in logical thinking. Of course many of Einstein's thought experiments can now be carried out (and some were within his lifetime) but the habit of using them remains the standard method of teaching and understanding Special Relativity. For our first one, we picture an everyday situation. We are driving along in a car at a constant speed of fifty miles per hour. Our speed is, of course, measured relative to the ground. In front of us is another car, travelling at fifty five miles per hour, also relative to the ground. That is, it is going five miles per hour faster than our car, so to us it seems that the car in front is travelling away from us at five miles per hour. Hence that is its speed relative to us.

Similarly, to the occupants of the car in front, our car appears to be moving away from them at five miles per hour, in the opposite direction. As Einstein might have said, that is relativity.

Galileo identified this type of situation as leading to a general principle of relativity, which can be stated as follows:

> The measured speed of anything will depend
> on what you measure its speed relative to.

It has to be conceded that this statement is not the most elegant use of English. It also has to be admitted that it is a very simple and naive statement of Galileo's relativity, and that we shall have to develop it into a more formal and general principle as we go on. Nevertheless, it is a perfectly correct statement and will do for the moment.

We can now give a preliminary answer the question posed in the title of this book. What is so 'special' about Einstein's Special Relativity? Well, Einstein agreed with Galileo's principle of relativity up to a point. But, according to Einstein the speed of light in vacuo was an exception to Galileo's general rule. Einstein postulated that the speed of light was an absolute quantity and did not depend on what you measured it relative to. Now this was seen by physicists at the time as surprising, even shocking. We will discuss why this was so in a later chapter. At this stage we note that this simple postulate led to all the bizarre ideas that we associate with Special Relativity. In particular, the result was that one could now say the following things about any moving body when compared to that body at rest:

1. A moving body contracts (i.e. becomes shorter) in its direction of motion.

2. Time passes more slowly on a moving body (This is the so-called time-dilation effect and leads on to the famous 'twins paradox').

3. The mass of a body is greater when it is moving, than when it is at rest.

It should be emphasised that these effects are only appreciable for speeds approaching that of light. So for most purposes, the picture due to Galileo (and Newton) is perfectly adequate. However, perhaps the most famous result of Special Relativity is the mass-energy equivalence. This has been described as the most famous equation in physics and is normally written as:

$$E = mc^2.$$

Or, in words, 'the energy of mass m is equal to the mass multiplied by the square of the speed of light'.

Before we leave this short, preliminary treatment of the topic of relativity, we should consider a general point about the interlocking nature of physics. When we mentioned the relationship of Einstein's work to that of Galileo, we used the word 'correction'.

I want to make it clear that the emphasis here is on the word 'correction', Often, people refer to the 'overthrow' of Newton's laws by Einstein's work. But this, although understandable, is rather misleading. In fact such a reaction was only really justifiable at the time Einstein published his work, as things did not stop there in 1905. A continuous process of scholarship established that Einstein's theory formed a continuum with the Newtonian picture, with a smooth transition from one to the other as the speed of a body increased towards that of light. The same considerations apply to similar exaggerated remarks about quantum mechanics versus classical mechanics

From time to time sensationalist comments are made about claims of particles moving faster than the speed of light. The most recent of these that I know of was in 2012, when some results from the Large-Hadron Collider at CERN were supposed to indicate neutrinos travelling faster than light. Such an event, if true, would not be 'too bad for Einstein' (sic)! It would in fact be 'too bad' for the whole interlocking structure of modern physics, and we could say goodbye to all the electronic conveniences of modern life, from smart phones through super reliable transport to life-saving medical advances, and much besides. All of these things depend on modern physics.

We shall enlarge on this aspect in a later chapter. In the next chapter, we consider another aspect of the nature of physics which we shall need in order to go into the subject of relativity in greater depth.

Chapter 2

How do we say what we mean in physics?

There is an old adage, 'Say what you mean and mean what you say'. In physics the second part can pretty well be taken for granted: everything is subject to verification, and dishonesty is rare. So the emphasis is on the first part, and in fact the need to do this lies at the heart of physics. We have to try to be as precise as possible, and avoid any ambiguity. Of course, physics is not unique in this respect. We could, for instance, draw an analogy with the drafting of parliamentary legislation, where the need to avoid ambiguity and hence misunderstanding is paramount. Unfortunately, just as legal language can seem turgid and obscure to those who are not lawyers, physics can seem pedantic, and this is off-putting to the non-physicist.

While we will make it as accessible as possible, we

do have to make use of the precise terminology of relativity theory. We will do this increasingly as the book goes on, but here we will introduce some of the key terms. As you will quickly see, they are are quite simple, and repeated use will make them familiar.

We begin with the idea of an event. This has the same meaning as in everyday life. It is a happening: something which happens at a specific place and time. If we wished to be formal, we could say that an event is represented by a point in the space-time continuum. As examples of events in everyday life, we could have something major like an election, something minor like a family wedding, or something in-between like a concert. However, in physics an event might just be someone switching on a light or taking a reading from a thermometer.

This brings us on to our next concept: that of the observer or experimentalist. In relativity, we deal with the activities of observers. Taking or making an observation is not just a question of an individual looking at something. That would be too vague. In relativity we are always concerned with an experiment where a precisely defined action is carried out or a reading is taken or a measurement made.

Our third concept in the study of relativity is that of the reference frame which is often referred to as 'an inertial frame'. The idea of a reference frame (or, more usually, 'frame of reference') is perfectly familiar in a non-scientific sense, and indeed is often used as a metaphor. However, in physics, as in so many other fields, such as surveying, planning, navigation, design, map-making, and so on, it is a system for taking measurements. For instance, we can obtain the distances between places from a map, these distances having previously been determined by surveyors using an agreed convention. The map is a pictorial representation of our reference frame and, if it should be (say) a map of Scotland, then Scotland would be our reference frame.

A frame of reference can be anything: the Earth, the solar system, the fixed stars or even your sitting room. The only criterion for choosing any particular frame is that it should be suitable for our purpose. Obviously, if you are laying out your garden, then any of the above frames would be totally unsuitable. The most suitable frame in that case would probably be dictated by the boundaries of the garden.

For relativity, there is a more subtle constraint on our choice of reference frame. We require it to be a frame in which Newton's first law of motion holds. We will discuss Newton's laws in a later chapter but for the moment his first law (also referred to as N1) states that:

> A body continues in a state of rest or uniform
> motion unless acted upon by a force.

By 'uniform motion' we mean that a body travels in a straight line and at a constant speed. Also, a body can be anything possessing mass: a planet, a cricket ball or even a human body.

If, as I suspect it may, this statement of Newton's first law seems rather obvious to you, then that is a tribute to the progress that we have made from the Enlightenment onwards. When Newton first put it forward, it would have been seen as quite radical, in a context where religious philosophers were fully prepared to specify the precise number of horses needed to pull the sun in its chariot on its daily journey round the sky!

What it means in practice is that an inertial frame is one that is at rest or is moving with constant speed in a straight line. Accelerating frames are excluded, as are rotating frames. For instance, if we were to put a marble down on a rotating record turntable, then the marble would move radially outwards. Whereas if we were to place a marble on the floor of a train moving at a constant speed, then the marble would remain at rest. Of course if the train were to slow down or speed up, then the marble would move, and so that train would no longer be an inertial frame, until it was once again moving at constant speed.

The study of relativity involves (at least) two frames of reference. First, we have an experimentalist observing an event in a frame where both observer and event are at rest. This is called the 'rest frame', although nowadays in physics there is a growing preference to say 'the co-moving frame'. Secondly, we have an experimentalist who is at rest in a second frame which is moving relative to the first. This experimentalist observes the event as moving away from him. Clearly we have a choice in that we can think of the second observer as being a moving observer, or as a stationary observer making measurements on a moving event. Again, as Einstein supposedly said, that is relativity.

This may seem a rather abstract definition of the study of relativity, but we shall put some flesh on these dry bones presently, when we consider specific examples.

Another aspect of saying what we mean is the use of symbols. So far we have been using terms like 'five miles per hour'. From now on, we will write this as '5 mph'. This notation will be perfectly familiar to most readers, and it allows us to express ourselves in a more compact way. Ultimately it leads on to mathematics. This is the language of physics, but unfortunately for many people outside the world of science (and indeed some inside it!) the use of mathematics arouses alarm and despair. In this book we will have to try to get round this problem. We shall not carry out any actual mathematical manipulations, such as solving equations. But we would suggest that maths can be treated as a foreign language and, just as a Latin tag might be unknown to us but could be translated, so a mathematical equation can be made understandable by translating it into everyday language.

Chapter 3

The willing suspension of disbelief

Recently the lights in my house went out and I had to call in an electrician to make some repairs. In the course of our conversation he asked what I was a professor of. My reply that I was a professor of physics produced a fairly standard reaction, albeit in a rather extreme form. He told me that he had studied physics at college and that 'no normal person could understand it!' He also volunteered that 'it would take a very special kind of brain to understand physics.' I thought it ironic. As someone whose very job depended on applied physics, an electrician might surely be expected to be a little more knowledgeable about physics than the average person.

However, although his comments were meant to be jocular, there may well be some truth in them. There does seem to be something of a division into 'scientists' and 'non-scientists'. Many years ago, I read an account of an experiment carried out at an American university, in which physics students were persuaded to opt for an arts course for one semester; and *vice versa*. I have no memory of the reactions of the physics students but I do remember what the arts students had to say about physics. A comment that cropped up again and again was: 'My opinion was of no importance!' Also: 'There was one correct answer and you were expected to get it.'

One might have thought that they would have met such concepts at an earlier stage in their education: perhaps in arithmetic, for instance? Nevertheless, it does seem to me that there must be some kind of dichotomy involved, and this is a thought that has come back to me many times over the years. So, although I do not have a definite answer to the problem posed by the differences between learning about the humanities and the sciences (bearing in mind that there is a spectrum of activities in both), I do have some tentative ideas.

Let us compare the study of history with that of physics. This discussion is necessarily going to be greatly over-simplified, yet I think it may be possible to extract something helpful from it. In history, as in all disciplines, we begin by acquiring facts. And in history these facts are of a familiar type. Essentially, people are born, grow up, marry, have children, and ultimately die. I believe that this sequence of events is summed up by provincial newspapers as 'hatching, matching and dispatching'! So, trivially, this is the stuff of everyday life. Hence the study of history is firstly a matter of becoming familiar with these facts for some specific set of people. Then there is the question of interpretation. Suppose that a particular monarch had three daughters. Evidently, in the absence of sons, there may be a problem over succession. Or, perhaps the marriages of the daughters to other monarchs may bring yet more problems. This is where there is room for opinion and hence for disagreement. As the recent upsurge of interest in the Great War has shown, there can still be controversy over its origins, even although the basic facts are well known. Also, it is worth noting that adopting a revisionist approach in order to show an original angle on the subject may well be a temptation to the ambitious historian.

At first sight, in physics the situation could not be more different. Indeed, in both relativity and quantum physics, some concepts are not only unfamiliar but actually run counter to our everyday experience. And even in the more traditional subjects of classical physics there can be concepts which are not easily understood in terms of our everyday life. Most people (physicists included) find entropy (which crops up in the study of thermodynamics: see the endnotes for this chapter) to be a difficult concept.

So what is to be done to bridge the gap? I think that the responsibility for this is shared (although perhaps not equally!) by the author and the reader.

The author needs to introduce concepts gently and progressively, and to some extent attempt to domesticate them by drawing analogies to their everyday counterparts. He must also provide a narrative to keep things moving along. In short, he must tell a story.

At the same time, the reader also has a part to play, because this is inherent in the very act of reading, or of being entertained in any way. Traditionally, a story is seen to rely on the reader's 'willing suspension of disbelief'. The analogous step for physics is for the student to be patient, and not to be put off by the fact that things may not make complete sense right now. In other words, physics students probably approach their lectures with the tacit assumption that they are not always going to make complete sense straightaway. As I have indicated, the concepts of physics can seem esoteric. And perhaps in some cases familiarity has to be acquired as a step on the way to understanding.

As ever, the exceptions serve to prove the rule. In my experience, over the years there has been the occasional student who has not demonstrated this patience, and has put up their hand during a lecture and said 'I don't understand that'. The rest of the students who were quietly taking notes would look rather pityingly at the questioner, although possibly many of them felt much the same way about things. My problem was to respond without crushing the student, so I would ask 'Do you understand Mayfield Road?'

After the student had reacted with amusement or bemusement, or a mixture of both, I would explain. I would point out that when they first came to Edinburgh, they would walk up Mayfield Road from their halls of residence in the town to the Kings' Buildings science campus in the suburbs. At first the road would be strange to them, but in time it would become familiar. Once that had happened, they could then consider whether they understood it or not!

So to complete our analogy, the history student can be thought of as walking along a road which is perfectly familiar, but as not knowing at what point they will turn off into another road. In contrast, the physics student journeys along a road which is quite unfamiliar, and in addition also doesn't know where they will turn off.

Hence the deal is this. As the author, I will make the story as much like history as possible. And you the reader must promise not to be put off by the occasional unfamiliar term, but instead should behave like a physics student, and let your understanding develop.

Chapter 4

Galileo's World 1: the geocentric universe

There were two aspects of Galileo's world which seem to have led him quite naturally to his concept of relativity. First, the heretical concept of heliocentrism was in the air. Secondly the best available means of long-distance transport was by sailing ship. And of course the recently invented telescope helped too. In this, and the following chapters, we will argue that the very fact Galileo saw that the Earth went round the Sun, rather than the Sun around the Earth (a relativistic disagreement, if ever there was one!), was conceptually linked to his recognition of the concept of relativity. This is an example of the interlocking nature of physics. As we will see later on, a similar reasoning also applies to Newton's discovery of the laws of motion.

In order to understand how Galileo came up with

the idea of relativity we have to try to understand the world as he saw it. Philosophically, this was based on the doctrine of the geocentric universe as propounded by Ptolemy (88 - 168 AD). Ptolemy started from the assumption that the Earth was at the centre of everything and worked out the consequences in considerable and ingenious detail. If Ptolemy's idea could originally have been described as a theory, then by Galileo's time it had long had the status of a doctrine.

Let us now consider the Ptolemaic universe, at the same time keeping a relativistic picture in mind. We begin by noting that probably we have all experienced a moment of confusion, when we weren't sure whether or not we were moving. You know the sort of thing. You are sitting in a train which has been stopped at a station for some time and your attention has wandered. Perhaps you have become immersed in reading your newspaper or book. Then you look up and have a momentary sensation of surprise that you are moving. But you aren't. Another train is slowly going past and for a moment (rather like an optical illusion) you switch reference frames and think that it is your train that is moving. One can have the same experience in an airplane which is taxiing.

I once experienced that momentary confusion on a much larger scale. At the time, I owned a house which had its main windows facing due south, but also had one full-length upstairs window (at the end of the landing) facing east. One evening I glanced out and noticed a very bright star in the sky above my neighbour's house. An hour or two later, I looked out again, and saw that the star was much higher in the sky. However, as I focused on the star, I experienced that illusory change of reference frame and it seemed to me for one vertiginous moment that my neighbour's house had become lower as the Earth rotated towards the east! Of course, a moment later and all was back to normal. The Earth is our natural reference frame and we refer pretty much everything to it. Normally we are completely unconscious of its motion, whether about its own axis or in orbit round the sun. So it is easy to understand our remote ancestors' thinking that the Earth was the centre of everything. Indeed, later on, we will have to overcome such habits of thought ourselves, if we are to understand relativity. In the meantime, however, let us take a look at Ptolemy's science.

It is quite usual to take the Earth as a frame of reference, in which we specify the position of any object by means of two angles, corresponding to its latitude and longitude, along with its height above sea level. For simplicity, we may assume the Earth to be a perfect sphere (it isn't!) and in principle we measure all angles by taking radii from the centre. This idea is readily extended to the heavens, in that we imagine as reference frame a celestial sphere which is centred at the centre of the Earth and extends radially to infinity. Then we can specify the position of any star in the sky by means of two angles, although the question of how far away it is, presents more of a challenge. The importance of this concept was that an astronomer in Greece, for example, could make use of tables of stellar positions which had been drawn up in Babylon, as could an astronomer in Egypt.

Thus the use of the celestial sphere as a universal reference frame permits us to construct a universal description of the night sky. The only problem is to find ways of measuring angles and referring them to the centre of the Earth. However, this is just a matter of simple geometry (and trigonometry), and we won't go into it here.

At the same time, it is worth emphasising that, prior to the invention of the telescope and over many millennia, observations had to be taken using the naked eye. The only instrument was the quadrant, which was a device for measuring angles up to ninety degrees. Quadrants ranged in size from a device you could hold up in one hand to something so big that you could climb onto it. In all cases, the observer relied on aligning pinholes or cross-wires with the stellar object being studied.

It makes an interesting 'thought experiment' to put ourselves in the shoes of these ancient astronomers. Now that we're in a position to track and record the motion of stars across the sky, let's see how we get on. First, while it's still daylight, we notice that the sun rises in the east, reaches a highest point (the zenith) and then sets in the west. At night the moon does much the same sort of thing, but the stars divide into two categories. There are those which remain still - we call these the fixed stars - and those that move. These latter, wandering stars, were called planetoi or wanderers, by the Greeks, and from this we get our term planets.

Ptolemy took the view that the spherical Earth was the centre of the universe and that all the heavenly bodies revolved round it. They were believed to move in circles, because that was a 'perfect shape'. However, when the orbits of the planets are plotted out, it turns out that sometimes they move backwards. That is, in an easterly direction. This is known as retrograde motion and is hard to explain on an assumption that the planets simply revolve round the Earth in circular paths.

Ptolemy was equal to this challenge and postulated that the planets also moved in small circles, known as *epicycles*, which were centred on their big orbital circles. Then, as the centre of an planet's epicycle was moving westward, the planet itself could be moving eastward for a while. Other problems needed other assumptions; but, as the model could be made to fit the observations, and even had predictive power, it seemed pretty satisfactory. So much so, that it remained the preferred theory until the invention of the telescope permitted observations which were to rule it out completely. However, even before that time, the multiplicity of modifications needed, ensured that it tended to run foul of Occam's Razor, making it vulnerable to the formulation of a simpler theory. For further discussion of this point, see the chapter endnotes.

Chapter 5

Galileo's World 2: the heliocentric universe

The first serious challenge to the Ptolemaic universe came from Nicolaus Copernicus (1473 - 1543), who put the sun (helios, in Greek) at the centre of things. A man of many parts, he was a Catholic cleric whose main work 'De revolutionibus orbium coelestium' or 'On the Revolutions of the Celestial Spheres' was banned by the Catholic Church. His main fields of interest ranged over mathematics, medicine, and canon law, with astronomy being, in effect, a hobby. Yet, as we shall see, one observation was enough to trigger off his challenge to religious orthodoxy.

The thinking of the medieval Christian church in these matters was determined by the vision of the Greeks, as transmitted, and further developed, by Arab-Islamic science. In short, natural philosophy was 'according to Aristotle', and the heavens were 'according to Ptolemy'. At that time, the idea that those heavenly bodies which move do so in circles around the Earth was seen as quite a natural one. In case you may be carried away by its naturalness, and tempted to accept this yourself, I should point out that the official Roman Catholic picture at that time also included the fact that the planets, Moon and Sun were carried round in their orbits in chariots which were pulled by winged horses!

So Copernicus, in questioning the Ptolemaic vision, was attacking a view which had prevailed for a millennium and a half. Apparently the inconsistency which aroused his interest, was that the brightness of Mars varied by such large amounts that it couldn't be accounted for by Ptolemy's epicycles. Also, when it came down to it, Ptolemy's theory had to be readjusted every couple of centuries or so, in order to minimize errors in the long-term predictions of planetary positions.

Then there was the aesthetic aspect. Copernicus thought that the Creator would naturally put the Sun, the source of both heat and light, at the centre of things. However, the main advantage of the heliocentric theory was that it explained retrograde motion of the planets in a natural way. If the Earth itself is moving round the Sun, then relative motion between Earth and a particular planet, can make that planet seem to 'go backwards' relative to the Earth. So, there was no need for Ptolemy's epicycles.

After the death of Copernicus in 1543, the concept of a heliocentric universe languished and more than a hundred years were to elapse before Galileo stuck his neck out, and asserted its truth. In the meantime, the next major player on the scene was the Danish astronomer Tycho Brahe, who was born in 1546, into a rich, powerful and aristocratic family.

Tycho was a colourful figure who wore a nose made of gold and silver to replace his own, which had been cut off in a duel when he was twenty. Apparently the quarrel was over a mathematical formula. In comparison, our modern custom of 'peer review' for settling academic disputes seems quite insipid! He made a morganatic marriage to the daughter of a Lutheran pastor. However, the strict Danish laws on inheritance in such a case ultimately led him to move to Prague, where things were a bit more liberal, and his numerous children could inherit his property. He died in Prague in 1601.

There are competing theories about his death. These range from natural causes (due to his being too polite to leave a banquet and hence straining his bladder) to mercury poisoning, either accidental or deliberate. In the latter case, there are again competing theories: one school of thought is that the King of Denmark sent someone to poison Tycho, while others think that his assistant, Johannes Kepler, murdered him. Believers in the former theory also believe that Tycho was the inspiration behind Shakespeare's Hamlet.

However, the really interesting thing about Tycho is that at the age of seventeen (while still a law student) he concluded that existing astronomical data was so inaccurate that it was impossible to distinguish between the competing theories of Ptolemy and Copernicus. Accordingly he set out to produce good observational data, and the observatory which was built for him by the then King of Denmark (not, presumably, the one who is supposed to have had him poisoned!) is believed to have cost more than five percent of Denmark's gross national product. Hard to imagine that happening nowadays. In this lavish observatory, Tycho built the largest astronomical instruments to date, including a quadrant with a radius of two metres. In modest contrast, Copernicus had his house built with a slot in one wall so that he could measure the time at which planets crossed the meridian.

Tycho is credited with observations of both a new star (a phenomenon now known as a supernova) and a comet. These events were of course incompatible with the traditional view of the heavens as being immutable and unchanging. So one might have supposed that this would persuade him towards the Copernican view. But apparently it didn't. He tried to work with a mixture of the two theories, preserving as far as possible the philosophical picture of Ptolemy (Earth-centred), along with the practical benefits of Copernicus (Sun-centred).

From our point of view, his achievements are twofold. First, he set out to rectify what he saw as an unsatisfactory situation. To do this, he amassed a huge amount of data in a systematic way. He also introduced the idea of the accuracy or uncertainty of an observation: this is a required feature of physics nowadays but at that time was new. Secondly, he gave us Kepler. And Kepler's laws of planetary motion are still used today.

Johannes Kepler, astronomer (and, it would seem,

murder suspect), worked obsessively over many years to use Tycho's comprehensive data to decide between the geocentric and heliocentric pictures. Although ultimately coming down in favour of the latter (and hence Copernicus), he found it impossible to reconcile the data with the idea of circular orbits. Accordingly, he tried other shapes and came up with the ellipse. He found that the astronomical data could be accounted for if all the planets moved round the Sun in elliptical orbits. From this result, he formulated his three laws (see the endnotes for this chapter).

Kepler's approach belongs to what we now call 'phenomenology'. Nowadays undergraduate students derive Kepler's laws in a purely theoretical way, using mathematics and the general principles that are laid down in Newton's laws. It is part of their elementary coursework in university physics and only involves a few hours work!

Chapter 6

Galileo's World 3: a near-death experience

In January 2008, Pope Benedict XVI cancelled a planned visit to *La Sapienza* University in Rome because staff and students were accusing him of having defended the Inquisition's treatment of Galileo. However, according to the Vatican, the Pontiff had been misquoted. Also, as they were busy erecting a statue to Galileo within the Vatican, clearly they had now forgiven his having been in the right in the first place, some four hundred years ago. Perhaps there were other, less celebrated, people such as Giordano Bruno, who also deserved a statue.

In 1560, Giordano Bruno was burned at the stake in Rome. His sin? He had advocated a heliocentric universe. At that time, Galileo was thirty six; and thirty years later he was to stand trial in Rome for the same sin, and with the same possible outcome! So, how did

Galileo, nothing if not forewarned, get himself into that situation? It certainly would seem to indicate, if not foolhardiness, a remarkable degree of confidence in himself.

To begin with, Galileo had the advantage of being quite well born and well connected. His father, Vincenzo Galilei, was a skilled lutenist and a musical theorist. He was somewhat ahead of his time in advocating the use of equal temperament: the twelve-toned modern musical scale. His success attracted powerful patrons, and he was eventually able to marry into a noble family. So Galileo had a good start in life and this would contribute to a feeling of self-confidence.

In his development as a physicist, he also had the advantage of his family's musical background. At that time, musical theory was one of the few really scientific subjects, and Galileo's father was the first to establish that the pitch of a note on a stringed instrument was proportional to the square-root of the tension applied to the string. Galileo himself is credited with being the first to recognise that the laws governing the natural world were a matter of mathematics rather than philosophy. Moreover, the mathematical structure of musical theory still has relevance to physics today. In fact, anyone wishing to understand quantum theory might be well advised to achieve an understanding of musical theory first.

Before turning to Galileo's famously controversial role in promoting the heresy of Copernicism, we will briefly summarize his achievements in physics. His most important contribution was to study the behaviour of falling bodies. The idea that he did this by dropping things off the Leaning Tower of Pisa is, alas, probably a myth. Rather more prosaically, he rolled metal balls down inclined planes. He established the mathematical form of the laws of motion under constant acceleration (due to gravity, in this case) and that the trajectory of a projectile is a parabola. This work laid the foundations for Newton's approach, which we will meet in a later chapter. He also observed the fact that a swinging pendulum had a constant period which depended on its length, but not on its amplitude. This idea was followed up by his son, who drew a plan for a pendulum clock. Although, the first actual working example of this was invented by Christiaan Huygens in 1656. The pendulum clock was the most accurate way of measuring time until the 1930s and of enormous importance in the development of physics, which relies on our being able to measure mass, length and time. However, what concerns us at this point is the invention of the telescope and the resulting revolution in astronomy.

The first spyglasses or telescopes (the latter term was coined by an associate of Galileo) were made by spectacle makers in the Netherlands (in particular, by Hans Lippershey) and are dated to 1608. They were made from standard spectacle lenses and were really just toys which quickly became wildly popular. Galileo was distracted from his existing scientific work by the arrival of one of these toys and immediately set to work developing his own practical telescope. This involved not only the theoretical step of working out the best type and combination of lenses, but also doing the actual lens grinding.

At first his telescope was used for purely terrestrial purposes, in particular to see ships while they were still far from shore.

This caused such interest that he was able to improve his university position. But, his position was to become totally transformed by the accident of his turning his telescope in the direction of the Moon. To his astonishment, he found that the Moon was like the Earth, with mountains and valleys. Other discoveries followed rapidly, including sunspots, the phases of Venus, the moons of Jupiter, and the fact that the planets were all the same kind of body as the Earth or Moon, whereas the fixed stars were like little suns. This was all contrary to the established view, going back to Aristotle, that the heavenly bodies were all perfect and unchanging spheres.

This was heady stuff and Galileo rapidly became a very famous man. He acquired powerful patrons and a reasonably well-paid appointment for life, with no onerous teaching duties. He adopted the Copernican view, but, in expressing this openly, he was, literally, playing with fire. The Vatican was not amused and Galileo received a written warning. Despite this he went further and published his book 'Dialogue on the Two Chief World Systems: Ptolemaic and Copernican' in 1632 and as a result found himself on trial for heresy in 1633.

Various factors would have played a part in his finding himself in this position; apart from the central fact that any scientist would find it very painful to be told what he can or cannot say. There may, for instance, have been an element of hubris. Any successful person may fall victim to their own publicity! And, the new Pope had been an admirer and acolyte (as regards science) of Galileo. Also, rather ingenuously, Galileo had adopted the expedient of writing out his ideas as a dialogue between two men, so that the opinions expressed did not necessarily represent his own views. In fact this was the basis of his first defence: that he had not said that the Earth moves round the Sun.

This didn't succeed and his revised defence, which was accepted, was that he had said it, but only inadvertently. He was sentenced to life imprisonment, which was commuted to house arrest, and required to make a formal recantation of his belief that the Earth moved round the Sun.

There is nowadays, in some quarters, a revisionist view to the effect that the Church was defending scientific values by condemning Galileo for asserting something which still lacked experimental proof. Certainly, as in most human affairs, there was a mixture of personalities, politics and matters of principle. But, one also senses a degree of vindictiveness in the way that Galileo was treated.

We shall not quote Galileo's abjuration of Copernicism here. Instead, we finish with the following words by him:

> I do not feel obliged to believe that the same God who has endowed us with sense, reason and intellect has intended us to forgo their use.

Evidently Galileo had a rather dry sense of humour. This remark could well serve as a motto for anyone who becomes involved in a dispute between science and religion.

Chapter 7

Galileo's law of the addition of velocities

Galileo's ideas about relativity started with his own experience. He remarked that, when in a closed cabin on a ship, he was often unable to tell whether or not the ship was moving. Or, for that matter, if moving, in which direction. Of course all he had to do was go up on deck and take a look. Then he would have seen whether or not the ship was moving relative to something else. In other words, if he saw the lighthouse slipping past on the port bow, it would be perverse to assume that the lighthouse was in fact moving and the ship was stationary. So motion only had meaning relative to something else. In this case, the harbour, the lighthouse or even the sea, would provide a fixed reference frame.

This inspired Galileo to carry out various thought experiments. He imagined studying the flight of gnats, or the swimming of fishes, or the behaviour of drops of water or of thrown objects. He concluded that none of these experiments, if carried out in his closed cabin, would tell him whether or not the ship was moving. Accordingly, we can state the principle of Galilean relativity as follows:

Galileo's principle of relativity is as follows:

> No mechanical experiment carried out in a closed cabin can be used to tell whether a ship is moving or at anchor.

Then if we wish to generalize this to something more modern-sounding we can convey much the same message in the following form:

Galilean relativity states:

> No mechanical experiment can be used to tell whether an inertial frame of reference is moving or at rest (with respect to any other frame).'

Now we recognize that what is true for a ship is true for any uniformly moving body which can be regarded as a frame of reference. This also applies to our own planet, but this fact can be difficult to see. Indeed, it can feel positively counter-intuitive, and over the years I have encountered the occasional student who was prepared to say so.

In any particular instance, I dealt with this as follows. I told the student to close his eyes. Then I asked him, was he moving? Naturally he would reply no, that he wasn't. Then I told him he could open his eyes again. When he did so, I would point out to him that he had in fact been moving. He had actually been moving at considerable speed with the Earth's rotation about its axis. Also, simultaneously he had been moving with great speed as part of the Earth's revolution around the Sun. As, of course, he still was, even with his eyes open.

In practice I found that students tended to be quite satisfied by all this and, as a result, were then prepared to accept this central principle of relativity. But it is instructive to consider just why the idea is so counter-intuitive. When we are in motion, it is usually relative to the Earth. Locally, the Earth makes a perfectly good inertial frame of reference. The point which I wished (and wish) to get across, is that we cannot detect uniform speed but we can detect acceleration. If our motion is accelerating, then we experience a force. For instance, there is that thump in the back you get as the motor car accelerates. Likewise, if the car is slowing down, we again experience a force. The problem with driving along at constant speed is that we are still subject to numerous forces in the form of vibrations (not to mention the noises which go with driving in a car). So even as a passenger, with eyes closed, you know when you are moving and when you are not.

In Galileo's case, the ship would have been a sailing ship and so fairly free from noise and vibration when under way. Also, his 'closed cabin' meant that there was no window or porthole. And lastly, the various horizontal and vertical accelerations due to wave motion would

41

have been present whether the ship was under way or at anchor. So possibly it was easier for Galileo to come to the correct conclusion.

However, his instinct was to get away from subjective impressions and actually do some experiments, and we could do worse than follow his example. But we won't think in terms of gnats or fishes. Instead, let us consider something a bit more scientific. For instance, let us assume that we have a simple pendulum suspended from the roof of a car and that our experiment is to measure its periodic time by means of a stopwatch. If the car is a rest, then the periodic time of the pendulum will be exactly the same as if the pendulum was suspended outside the car. Obviously!

However, what happens if the car is moving? Is the periodic time affected? The answer is that it is not. So long as the car is moving at a constant speed, however large of small, the periodic time will be unaffected.

But if the car is accelerating or braking, then the periodic time will be affected. This is because the periodic time is controlled by the gravitational acceleration g and the length of the pendulum. Now, if the pendulum is not actually swinging, then it will hang vertically from the roof, while the car is moving at constant speed. If the car accelerates or brakes then the pendulum, will hang at an angle to the vertical, under the joint influence of gravity and the car's acceleration. This means that it is subject to a modified acceleration which is the resultant of the gravitational (vertical) acceleration and the car's (horizontal) acceleration, thus giving a different periodic time when it is actually swinging. Incidentally, this behaviour of a pendulum is (or used to be) the basis of the measurement of the braking efficiency of a car in the motor industry.

So, what is true for the simple pendulum is true for any mechanical experiment. That is what Galileo saw, and to this day it is referred to as Galileo's relativity. Where Einstein made a contribution was in extending this principle to all of physics; and that in particular meant including the electromagnetic field, which was unknown in Galileo's time. Even more daring was Einstein's second axiom, as briefly mentioned in Chapter One, that the velocity of light was an absolute rather than a relative quantity. In order to see why physicists found it so shocking, we have to consider the topic of relative speeds. Fortunately this is entirely familiar to us in our everyday lives and therefore we can call upon our intuition to help us.

While, as I have just said, this is a familiar concept, in physics it is usually referred to as addition of velocities, rather than of speeds. The distinction between velocity and speed is that velocity is a vector (i.e. its direction matters) and speed isn't. In fact speed is the magnitude of velocity and is a scalar. For instance, if we specify that a ship is travelling at 15 knots in a north-easterly direction, we have specified its velocity, whereas 15 knots alone would just be its speed. It also applies to bodies that can only move along a particular straight line. In other words, bodies that are in one-dimensional motion. Even in this restricted case, we can still specify a direction. This is a matter of convention. We could, for instance, say that bodies moving from left to right have a positive velocity, while those moving from right to left have a negative velocity.

In everyday life, we know that if two cars, each travelling at 30 mph, collide head on; then, for either car it will be as if they had hit a brick wall at 60 mph. That is, we add the velocities, because the cars are going in opposite directions, and we must use vectors. Similarly, for two-dimensional motion, if we want to plot an interception course for a customs vessel wishing to check out a ship suspected of smuggling, then we must resort to vectors because of the angles involved.

However, if we consider only one-dimensional motion then we can use speeds rather than velocities. And, this is what we will often do, because it is so much simpler. At the same time it is not dumbing down, because this convention is widely used in studying Special Relativity, where it is part of what is known as the standard configuration. Problems in relativity can be so subtle and difficult that we need all the help we can get! But that won't concern us here, as we are going to stick to some ideas which are very simple yet are realistic enough to enable us to understand the basic physics.

In order to update Galileo, our basic thought experiment will be based on the travelator. This is named by analogy with the escalator (or moving staircase) and is a moving pavement or walkway. They are encountered in many of the world's larger airports. So they make a convenient everyday demonstration of relativity. For Galileo, relativity involved a comparison between a fixed frame (the Earth) and a moving frame (the ship). Our fixed frame is provided by the ground (or Earth) while the travelator is our moving frame.

Basically there are two ways to use a travelator. Either you stand still and allow it to carry you along or you walk on it in your normal way. If you do the latter, then you will be moving much faster relative to the ground. Furthermore, people who are walking in the opposite direction (On the ground, of course! We are not allowed to walk on a travelator against the direction of travel; and in any case to do so would be foolish and pointless.) appear to approach one so much more rapidly that the effect can be quite startling.

If you ignore the travelator, then you cover the ground at your normal walking speed. We can write this as an equation, thus:

My speed relative to the ground = My walking speed.

Now suppose you use the travelator, and walk on it at your normal speed. Then you will be moving along more quickly than if you were just walking on the ground. Again, we can write this as an equation:

My speed relative to the ground = My walking speed + The speed of the travelator relative to the ground.

Note that we have just established a profound truth in physics by using a simple everyday example. Now we generalise it. We consider the speed of anything (e.g. a person, a motor bike, an aeroplane, a sound wave, a planet, a molecule, a light wave, ...) relative to a moving frame (e.g. the travelator) and relative to a fixed frame (e.g. the ground). Then our general Galilean statement of relative speeds (or addition of velocities) becomes:

The speed of any object relative to the fixed frame = The speed of that object relative to the moving frame + The speed of the moving frame relative to the fixed frame.

That is, we always have to add on the speed of the moving frame (e.g. the travelator) to get the speed relative to the fixed frame (e.g. the ground). The moving and fixed frames can be anything you like. For example, the Earth could be a moving frame relative to the solar system which is taken as a fixed frame. This illustrates the fact that the fixed frame is just a convenient convention. Previously the Earth was the fixed frame, now it becomes the moving frame. Every frame is moving relative to something else. But the above relationship is now obvious and must always hold. There cannot be any exceptions ... or can there?

In fact, this is exactly where Einstein stuck his neck out. He said that there was one exception to Galileo's law of relative speeds. And that was an electromagnetic or light wave *in vacuo*. We will explain the significance of the restriction to the speed of light in a vacuum later on (in Chapter Ten). For the moment let us marvel at the strangeness of Einstein's proposal. For the special case of light waves, we rewrite Galileo's law of relative speeds as:

The speed of light *in vacuo* relative to the fixed frame = The speed of light *in vacuo* relative to the moving frame.

But where has the speed of the moving frame gone? This equation is a special case of the one above it, but can only be reconciled with it if the moving frame has zero speed. Despite that, Einstein was postulating that the relationship would hold for any speed of moving frame. No wonder that physicists were shocked by this statement!

In order to make sense of this, we have to abandon Galilean relativity for speeds near the speed of light. Or to put it rather more precisely: what this means is that we have to abandon Galilean relativity altogether (i.e. for all speeds) for a new form of relativity, which then reduces back to the Galilean form, for speeds much less than that of light. It turns out that this step involves a distortion of the space-time continuum, and time itself is no longer universal.

We close this chapter with the reflection that when Galileo muttered 'Yet it still moves' after recanting his heretical views, he must have been aware of a painful irony. The difference between his view and that of the church was, trivially, a mere matter of relativity. For example, if we consider the Moon orbiting the Earth, then to a man on the Moon the Earth appears to be going round the Moon, whereas to those of us on Earth the opposite is true. It would take someone right outside the Earth-Moon system, a Martian perhaps, to establish which version was true. In a sense the telescope allowed Galileo to do just that, by seeing moons orbiting other planets.

However, in pedagogical relativity, this idea of going outside is forbidden. We have to restrict ourselves to what we can know within our system of inertial frames.

Chapter 8

Newton's laws of motion

A recent biographer of Isaac Newton (1642 - 1727) described him as 'The Last Sorcerer' (see the endnotes for this chapter). But, by this he did not mean, as people have said about the late Richard Feynmann, that he produced his theories in a magical way. He meant that Newton quite literally believed in magic. Of course other educated men at that time, including some of the great scientists, dabbled in alchemy. Yet somehow it seems rather incongruous that such a man, possibly the greatest rational mind of all time, should do so. Moreover, it was more than mere dabbling. It was a consuming passion which coexisted with his science and ultimately superseded it.

What also gives it a strong flavour of the paradoxical is that Newton was a one-man scientific revolution. Where his predecessors gave us increasing precision of measurement and the need to reconcile theory with experiment, Newton gave us an exact understanding of the solar system and our place within it. In the process, he ushered in the Enlightenment and the age of rational thought, which is still (I trust) with us today.

By all accounts Isaac Newton was not a pleasant man. If we overlook his carelessness in dress and (probably) personal hygiene, this unpleasantness ranged from minor matters, such as reserve and coldness, to major, such as vindictiveness, spite, ill-temper and positively dishonourable behaviour towards those whom he saw as rivals. He was quite unscrupulous in ensuring that the work of others did not seem to detract from his own. The astronomer John Flamsteed described him as 'insidious, ambitious, excessively covetous of praise, and impatient of contradiction ... a good man at bottom but, through his nature, suspicious.'

Yet, it was hardly surprising. He was born into a Lincolnshire farming family (which is probably a good enough start in life), but his father died soon after; and his mother remarried when he was three, going away to live with her new husband and leaving young Isaac to the care of grandparents. There seems to be no suggestion of his having been treated harshly, let alone with any deliberate cruelty. But betrayal and a lack of love is hardly a good start for any child. When his mother returned, after the death of her second husband, she opposed his desire to be a scholar and wanted him to follow his uneducated, yeoman father in working on the farm.

Fortunately, his talents were recognised both by his schoolmasters and by more academic members of his mother's family, and he ended up at Cambridge. His time there was probably less happy than it could have been. Due to his family's meanness over money, Newton had to take a semi-menial position in which he acted as a servant to more highly-born students. Given that his family were quite prosperous, and that Newton himself was later to inherit money, it is tempting to assume that his vindictive and spiteful qualities may have been inherited.

It is well known that Newton took a mediocre degree. Yet despite this he gained a college fellowship and later the Lucasian Chair of Physics. An interesting sidelight on his nature is that, at the age of twenty three, after being appointed to his college fellowship, he registered his entitlement to the rank of gentleman, and thereafter styled himself accordingly.

In the summer of 1665, the plague arrived in London and within a few months at least one in ten Londoners had died of it. Because of its proximity to London, the University of Cambridge was closed, and the students were sent home. Newton went home to Woolsthorpe Manor and, during the enforced idleness of the next eighteen months, he conceived most of the major ideas which were to be the basis of his scientific work over the next two decades. In the process, he brought to an end the Aristotlean view of physical science which had dominated natural philosophy for nearly two millennia.

Greek natural philosophy had been dominated by the work of Aristotle (384 - 322 BC) and, in the process, the much more intelligent views of the Atomists had been eclipsed. They were not to be resurrected until more modern times (see the endnotes for this chapter), when developments in physics and chemistry pointed to an atomistic view. Aristotle's philosophical position had been adopted without question by the mediaeval Christian church, and so had passed into the religious dogma which was to control scientific thought in Western Europe for centuries.

Aristotle's four elements of Earth, Wind, Water and Fire, now seem very naive, but his views on what we would now call dynamics were perhaps less so. Essentially Aristotle said that bodies remain at rest unless a force acts on them. In order for a body to keep moving, a force had to keep acting on it. For instance, an arrow in flight was supposed to be pushed along by the air rushing round behind it and pushing it forward. Of course, this explanation is readily seen to be incorrect: an arrow could fly in a vacuum. But Aristotle didn't believe that a vacuum could exist.

Nevertheless, Aristotle's views on this matter can seem quite sensible at first sight. Supposing I look round my study, what do I see? Well my writing desk appears to be at rest. As is the clock on the wall, the books in the bookcase and so on. Everything is pretty much at rest. However, if I push the desk it will move. So Aristotle's views on this matter seem to correspond to what we observe in everyday life.

However, there is a different way of looking at this. Suppose I were to claim that my desk was in motion as was the rest of the room. It is just that within the reference frame provided by the room (or by my house) that nothing is moving. But in fact everything is moving because the whole house is moving. As is the ground beneath my feet. My particular spot on the Earth's surface is rotating round the Earth's axis and has a corresponding linear velocity. And of course as the Earth orbits round the sun, my study has a pretty high velocity from this cause too.

So in one frame of reference (the Earth), everything in my study is at rest, while in some other frame (for instance, one could have a frame fixed to the Sun), everything in my study is moving, along with my house and indeed the ground underneath it. This fact was a key realization by Newton and led not only to his laws of motion, but also ultimately to Einstein's relativity. Newton's laws were celebrated in a famous couplet by Alexander Pope (1688-1744): 'Nature and Nature's laws lay hid in Night: God said, "Let Newton be!" and all was light.'

There are various reasons why the natural laws of motion lay hidden, but the major one is the existence of friction. If we set something into motion, it ultimately comes to rest because there are forces resisting its motion. We refer collectively to such forces as friction. It was for this reason that Newton spent a great deal of time studying resistance to motion, and particularly, air resistance. The property of fluid resistance to motion (be it air or water or any other fluid) is known as viscosity. To this day we use Newton's law of viscosity to define (and hence measure) the fluid viscosity. Fluids which satisfy this law are routinely described as 'Newtonian'. Exotic fluids such as nondrip paint, and many body fluids, are characterised by more complicated relationships and are known as 'non-Newtonian'.

It is this all-pervading resistance to motion which deceived Aristotle. We have the advantage of him, as did Newton, in that we realize there is no air resistance to affect planets and moons. So, like Newton, we can now understand the real laws of motion. We begin by summarizing Newton's laws of motion, just as we would in a textbook. There is a useful shorthand in physics which consists of referring to Newton's laws as N1, N2 and N3; and that is what we shall do here. We may set them out as follows:

N1 A body continues in a state of rest or uniform motion unless acted upon by a force.

N2 The rate at which the momentum of a body changes with time is proportional to the force acting on that body.

N3 To every action on a body, there is an equal and opposite reaction.

Now we shall look at each law individually in turn and try to see how Newton might have deduced it.

We already have all the pieces of the jigsaw from our previous discussions. All we need to know is that astronomical bodies, such as planets and moons, move endlessly through a vacuum and only deviate from straight-line motion because of gravitational forces. As we have noted, Aristotle was hindered from reaching this conclusion because he didn't think that a vacuum was possible. Newton realized that it had to be possible. And in later times evidence would come along to support this, ranging from measurements of the density of the atmosphere, which reduces the higher we go, to the invention of vacuum pumps, which could produce a perfectly good vacuum here on Earth. Of course nowadays we are all familiar with the fact that a space craft launched from Earth will travel on indefinitely out of the solar system, unless some other object gets in its way.

So the motionlessness of things around us is entirely compatible with Newton's first law. They have been brought to rest by frictional forces and will only be in motion when acted upon by some force which overcomes the friction. Thus in our everyday setting, Aristotle seems to be right. But, if we take the friction away, then once a body is moving, it keeps moving. An interesting aspect is that this also applies to the molecules of a gas, which move in a vacuum, as well as do the planets. So, in order to really understand our everyday world, where everything is on our scale, we find it helpful to understand what happens at very small (molecular) scales and at very large (astronomical) scales.

Another interesting aspect of N1, is that it is essentially a statement of conservation of momentum. We all know what momentum is in a general sort of way, but in physics it is necessary to be more precise. Let us do a thought experiment. Suppose you were playing rugby. Which would you prefer: to be tackled by a heavy or a light player? Or, by a player who was running quickly or slowly, just before the impact? Obviously you would choose the opponent who is light and running slowly; or, in other words, the one who has least momentum! This is because you realise that as the tackle takes place, you will experience their sudden change of momentum as a force. (Which means that you are aware of N2 as well.)

So we arrive quite intuitively at our definition of momentum. Remembering that the weight is just our everyday measure of the mass, or amount of matter, in a body we can define momentum as follows:

$$momentum = mass \times speed.$$

So, the greater the mass and the greater the speed of a body, the greater its momentum.

The first law can then be seen as stating the conservation of momentum and we can rewrite this as such:

The momentum of a body is conserved unless a force acts on it.

By conserved, we just mean that it remains constant. And it is worth remarking that conservation laws, such as those of energy, momentum and mass, are seen by many physicists as being the most fundamental way to describe the natural world. This is because they arise as a consequence of some underlying symmetry. Whereas, Newton's second law (which we are just about to discuss) is a purely phenomenological observation, as is Einstein's later version, and can be overturned at any time by a contrary observation.

Most people probably think of Newton's second law as saying something like:

The force on a body equals its mass times its acceleration.

Which is not exactly the same as my statement of N2 in the list of Newton's laws of motion given above. Actually they are the same, provided the mass of the body doesn't change; and in most situations that would be the case. But there are instances where the momentum of a body changes because its mass changes. For example, a falling raindrop or hailstone can increase or decrease its mass as it falls, depending on the atmospheric conditions. For the moment we will just note the possibility that the mass can change as something to be borne in mind.

So let's concentrate for the moment on acceleration. This is just the rate of change of speed with time. Nowadays we all have a huge advantage over Newton and his fellow scientists, in that we live in a world where personal experience of acceleration, and its possible relationship to force, is a matter of everyday experience. We have all experienced the 'push in the back' as a motor car or aeroplane accelerates. So the idea the force is just a product of mass times acceleration seems perfectly reasonable. If we extend our previous discussion about momentum, when being tackled at rugby by an opponent, then the faster our opponent is moving, the greater will be their acceleration (which is negative, of course, as they will be slowing down) when they hit us. And so will be the force with which they hit us. Also, the heavier our opponent is, the greater the force.

It is worth remarking that every day we are constantly subject to small accelerations (and their concomitant forces) in ways that we barely even notice. Walking, riding a bike (let alone a horse), on public transport, and so on, we are constantly shaken and jogged by small irregularities. Rather as Hamlet said, in his greatest soliloquy, 'the thousand natural shocks that flesh is heir to'. Although, I think that Shakespeare was referring to emotional shocks, rather than the experience of being shoogled about in a horse-drawn carriage. My reason for remarking on this fact is that it is so everyday a matter that it passes beneath our notice. Yet, I believe that it is the underlying reason that ideas of relativity (which we shall come back to in later chapters) are often perceived as being somewhat counter-intuitive. In other words, when we are in motion we are generally all too conscious of that fact!

Newton's third law seems to be very simple (and perhaps even obvious). But actually it can prove quite subtle. Let us begin with the easiest case. Supposing you engage in a pushing contest with someone and neither of you can budge the other. Clearly you are exerting the same force on your opponent that they are exerting on you; and in the opposite direction. Now, replace your human opponent with a wall (for safety's sake, choose a robust one!) and start pushing. It doesn't matter how hard you push, the wall pushes back and you are in static equilibrium. If you push hard you experience a strong reaction; if you push gently, you experience a weak reaction. But whatever you do, the wall pushes back in accordance with Newton's third law.

The most obviously interesting aspect of Newton's laws is the way in which they can be used to derive the equation of motion of a body subject to forces, with applications ranging from the prediction of planetary orbits to the motion of molecules. However, they also make important contributions to our understanding and it is this aspect with which we begin.

In physics we always try to work with the simplest situation which is compatible with what we are trying to achieve. So we will restrict our attention to the dynamics of bodies which have constant mass. Then N2 takes the form:

N2: force = mass x acceleration,

where everything refers to the particular body (for example, a molecule or a cricket ball or a planet) which we are considering.

We use this equation in the same way as a formula is always used. We put in numbers for the mass and the acceleration on the right hand side and do the multiplication in order to work out a value for the force on the left hand side. But there is more to it than that. Up until now we have been relying on our intuition about what a force is. We know that if we give something a push, then we are exerting a force on it. Now things change. We can use the above equation to define what we mean by a force. And, as we can measure the two quantities on the right hand side, this formula which we call N2 gives us a method of quantifying (or measuring) a force.

With the definition of force taken care of, the next thing is to use N2 as an equation of motion. In other words, given a body of a certain mass, and the details of the force acting on it, we can work out the acceleration of the particle and hence predict where it will be at any given time. To do this, we just swap over the two sides of the equation that is N2 and divide across by the mass, so:

$$\text{N2: acceleration} = \text{force} / \text{mass}$$

Now if we know the force (it could be gravity, for instance) we can predict what is going to happen to the body. In fact this is how the planetary orbits were calculated: by choosing the force to be Newton's inverse square law of gravity.

To close, we should point out that N1 and N3 are qualitative, and inherently and obviously true. In contrast, N2 is quantitative and, of Newton's laws of motion, it is the only one which may be overturned by a later form, such as that put forward by Einstein. We will come back to this point later, when we discuss Einstein's theories.

Chapter 9

How Maxwell and Newton jointly made Einstein necessary

Isaac Newton, James Clerk Maxwell and Albert Einstein make up the holy trinity of physics. Like every physicist whom we shall mention in this book, each of them was an outstanding individual. That almost goes without saying. In addition they were all prodigies, in the sense that Shakespeare or Mozart were. However, their position in physics does not derive from their status as individuals, but rather from their pivotal roles as unifiers.

As we have seen, Newton's work unified our understanding of the motion of bodies under the action of forces. In this context, Maxwell's contribution (in addition to his fundamental work on statistical physics and thermodynamics) was a unification of the laws of electricity and magnetism, now known as Maxwell's Equations.

James Clerk Maxwell has been curiously underval-
ued in his native Scotland. In Scottish histories or in lists
of Scotia's famous sons (and daughters) he is either miss-
ing or at best receives a brief mention. In the latter case,
this can be misleading, such as stating that he invented
colour photography. Strictly speaking, he didn't: what
he did was to demonstrate its feasibility. However, even
if he had, it would be quite a minor achievement compared to his real scientific work.

All of this is rather strange because in physics he is
one of the giants. Indeed, it is his work which made Einstein's Special Theory of Relativity necessary, and Einstein was said to have kept a photograph of Maxwell in
his study, along with portraits of Newton and Faraday.

Possibly one reason for this neglect is that, to the
non-scientist, Maxwell's work may seem rather esoteric.
Newton's work, however little people understood it, clearly
related to everyday matters (such as games of billiards)
and at the same time addressed great fundamental ques-
tions as to how the planets went round the Sun, or the
Moon round the Earth. Einstein's work, probably understood even less by the general public, at least had ex-
citingly bizarre consequences, such as time dilation and
length contraction. By comparison the fact that Maxwell's
work led on to the development of radio and television
(not to forget colour photography!), for instance, seems
rather prosaic and, well, technological by comparison.

James Clerk Maxwell was born in Edinburgh in 1831
into a family which ranked as middle class, being mi-

nor lairds and also financially comfortable. His parents were unconventional, having married unusually late in life, and his father, although an advocate, preferred the life of a country gentleman and inventor to the practice of his profession in Edinburgh. When young James ultimately attended school at the Edinburgh Academy (boarding with his aunt in Edinburgh during the week) his eccentric homemade clothing, which apparently had been designed on the same principles as a house, complete with rain gutters and drainpipes, combined with his rustic accent to make him a figure of fun to his fellow students.

Like so many of the great scientists, Maxwell showed an intense interest in understanding the world about him from an early age. Anything which moved, made a noise or in any way did something, would provoke the question: 'Wha's the go o' that?'. And he was not to be fobbed off. If an answer was not to his satisfaction, he would persist with: 'But, wha's the particular go o' that?'. He also showed a keen interest in making things and doing things for himself: a classic predictor of scientific creativity. While still at school he wrote mathematical papers which were presented to the Royal Society of Edinburgh, although someone else had to read them on his behalf as 'it was not thought proper for a boy in a round jacket to mount the rostrum there'.

James Clerk Maxwell had an outstanding academic career, first studying at Edinburgh and then at Cambridge University. He held university chairs in Natural Philosophy (as Physics was then known) at Marischal College, Aberdeen and at King's College, London. He then retired to work at his family home for some years before taking up his last appointment in 1871 as the first Cavendish Professor of Physics at Cambridge, where he was put in charge of the development of the new Cavendish Laboratory. He died of abdominal cancer in Cambridge in 1879, aged 48 and curiously at much the same age and from the same cause as his mother had done when he was a child of eight.

Before we turn to Maxwell's electromagnetic theory, there is one small aspect of his career which is worth mentioning. In 1860, the University of Aberdeen was formed from a merger of Marischal College and King's College (not to be confused with King's College, London or indeed King's College anywhere else) and the result was that one of the greatest physicists who ever lived was surplus to requirements and lost his job! He then lost out on the newly vacant chair at Edinburgh to his friend Peter Guthrie Tait, whose name is still commemorated today in the name of the chair of mathematical physics at that university. Still, after these minor setbacks, the rest of his career was a triumph.

Just as one speaks of 'Newton's laws, meaning his 'laws of motion', it is usual to refer to 'Maxwell's equations, meaning his 'equations of electromagnetism'. Also, just as in the case of Newton's laws we are concerned with the rate of change of velocity with time (i.e. the acceleration) and its relationship to forces, so with Maxwell's equations we are concerned with the rate of change of magnetic and electric fields with time, and the relationship to electromagnetic forces. As we do not 'do math' in this book, that restricts what we can say when discussing a set of equations. So we will have to see if we can get the general idea using words. It will be helpful to start with a simple idea about electricity and magnetism, as follows.

We know that electric charge comes in two forms: positive and negative. We also know that positive and negative charges attract each other and if they get together they cancel out. We call the result electrical neutrality. Now suppose we take a long rod of some insulating material, such as glass for instance, and put a positive charge on one end and an equivalent negative charge on the other. The result is what we call a dipole. If we plotted lines of electrical force, they would appear to come out of one end (or 'pole') and to go into the other.

At this point you are probably thinking to yourself 'This sounds a bit like a magnet.' And you would be right. It is rather like a magnet. Clearly one can draw an analogy between the positive and negative poles, on the one hand, and the north and south poles on the other. But, what is interesting now, is how our electrical dipole differs from a magnet in the following respect. If we break the glass rod into two pieces, then once again we have the two isolated charges, one positive and the other negative. In other words, the two ends of our electrical dipole have now become electrical monopoles. But, so far as is known, there is no such thing as a magnetic monopole. If you keep cutting a bar magnet into halves *ad infinitum*, you will always have one north pole and one south pole on each piece, however small.

So when we say that electricity and magnetism are really just aspects of the same thing (and later on we shall find that Relativity tells us that they are readily transformed into each other) it must be always borne in mind that in their basic phenomenology they are actually quite different. With this fact firmly in mind, we are ready to tackle Maxwell's equations, which can be stated non-mathematically as follows:

1. Gauss's theorem applied to an electric field.

   Gauss's theorem provides a relationship between the electrical charges in a system and the resulting electrical potential (or voltage) and hence, by simple maths, the electrical field. Technically we say that the presence of electric charges causes the electric field to diverge.

2. Gauss's theorem applied to a magnetic field.

   As we have noted, the magnetic analogue of charge - the monopole - does not exist. Accordingly for magnetic fields we conclude that there is no field divergence.

3. Faraday's law of electromagnetic induction.

   This is normally discussed in terms of induction coils, which is rather practical and quite specific. Maxwell's form is the general statement that rotation of the electric field lines is caused by a magnetic field varying in time.

4. Ampere's law of magnetomagnetic force.

   This deals with the force exerted between conductors carrying electric currents. Maxwell's general mathematical form of it is equivalent to the statement that the rotation of magnetic field lines is controlled by electric currents (i.e. the variation of electric charge with time)..

However, there is a complication here in that by 'current' we no longer mean the flow of charge through a conductor. When Maxwell attempted to write down Ampere's law in a general form, which would be capable of being applied to any situation, he ran into a mathematical problem. In effect, an inconsistency. He got round this by a purely physical assumption. He generalized the term 'current' to include a new current which he called the 'displacement current'. This allowed the effect of an electric field which was varying with time in an insulator to be included in the formulation.

At this point you may feel that Maxwell just 'adopted' a lot of other people's results. As the old saying has it: 'To steal the work of one man is called plagiarism. To steal the work of many men is called scholarship'. However, in fact Maxwell's contribution was threefold. First, he gathered together the four laws and recognized that they provided a complete description of the electromagnetic field. Secondly, he gave them a general mathematical formulation, which extended them to all possible situations, rather than just the specific situations from which they had been deduced in the laboratory. And, thirdly, he solved the technical, mathematical problem of formulating the fourth equation from Ampere's law which required the introduction of his displacement current. This requirement came from the pure mathematics and not from experiment.

In assessing Maxwell's equations, we should note that the first one stands on its own, irrespective of Maxwell, and is of tremendous practical importance. In practice, for static situations (that is, where the electric field does not change with time), one uses Gauss's theorem to derive Poisson's equation, and this can be solved for a given distribution of charges to obtain the 'contour map' of the potentials. Then the gradients of the potentials tell us the field. That is, where the lines of equipotential are close together, the electrical field is strongest. In other words, this is just like in a contour map, where the lines of equal height are packed together then the gradient is steep and the downward force acting on you is strong! This theory is used widely in applied physics, electrical engineering, electronics, astronomy, electrochemistry and many more. Accordingly, in the interests of completeness, it merits this specific mention.

Now we turn to electrodynamics, and in this context the first two Maxwell's equations are just conditions that the calculated electric and magnetic fields must satisfy. In the language of mathematics, they are called 'constraints. The key equations therefore, are the third and fourth. Essentially, these connect the spatial variation of the electric field to the time variation of the magnetic field, on the one hand, and the spatial variation of the magnetic field to the temporal variation of the electric field on the other. Technically, they are what is known as 'partial differential equations', but from our point of view what matters is that they are 'simultaneous equations'.

Now, even non-mathematicians will remember simultaneous equations from schooldays. We want to find the values of two unknowns, for instance x and y, when they are connected together by two simultaneous equations. The general method, you will recall, is to use one equation to form an equation for y (say) and then substitute that relationship into the other equation in order to eliminate y altogether, thus leaving one equation for the unknown x. Solving this for x, one then went back to the preceding step and substituted for x in order to obtain a value for y. In this way one obtained values for both unknowns, and the problem was solved.

And this, give or take a mathematical trick or two, is what Maxwell did with his two simultaneous equations for the electric and magnetic fields. He obtained a partial differential equation for the electric field which did not contain the magnetic field, along with a partial differential equation for the electric field which did not contain the magnetic field. The fascinating outcome was that the two equations were identical in form and that this form was what was called the 'wave equation'. In order to understand what this meant, we now need to make a slight digression on the subject of vibrations and waves.

The topic of vibrations and waves is a specialist area of physics which is relevant to almost every other aspect of physics, along with engineering and indeed everyday life. It underpins optics, the theory of light, radio and radar, water waves, the theory of music, structural engineering, the theory of stability, ... and so it goes on. It is a key subject.

Basically, a vibration is an oscillatory motion which is localised in one place while a wave is an oscillatory motion which is extended in space. As an example, take a guitar string. We pluck the string, and each point of the string vibrates back and forth (in what is known as simple harmonic motion) at right angles to the string. However, the amplitude of that vibration can travel back and forward along the string, between its fixed ends, and this motion we perceive as a wave.

Most waves are like this: their vibration is at right angles to their direction of propagation, and we call these transverse waves. Think of sea waves, for instance, or ripples on the surface of a pond, or children making waves along a skipping rope. However, sound waves, shock waves and the waves associated with earthquakes involve vibrations in the direction of propagation. We call these longitudinal waves. These are a bit harder to envisage but fortunately we do not need them here.

Sticking to the simple example of a vibrating guitar string, how would we go about developing a theory of the wave motion of the string? This is quite simple and is a routine part of first-year physics at university

When we pluck a guitar string, we displace it from its 'equilibrium position. Then we focus on one little piece of the string. We choose its length and hence we know its mass. At the same time, we know that the tension in the string provides the force which tends to bring the 'bit' of string back to the equilibrium position. All we have to do then is to invoke Newton's second law, and equate 'mass times acceleration' to the tension force and we can work out the acceleration of the elementary bit of string. Extending this analysis bit by bit to the entire string we end up with an equation of motion for the string: this equation is called the 'wave equation'. It describes sinusoidal transverse motion of the string in space and time and it predicts the velocity of the waves in terms of the tension in the string and the mass (strictly, the mass per unit length) of string.

I cannot emphasise too strongly that the wave equation is a universal equation which is found in all situations in physics where there is sinusoidal oscillatory motion And in all cases, it contains within it an expression for the wave velocity, in terms of a force-type element and a mass-type element.

Now we are in a position to appreciate what Maxwell found. The most general equation for the electric field was the same as the most general equation for the magnetic field and that equation was the wave equation. Moreover, the expression for the wave velocity was the same for both fields and depended on the magnetic permeability of free space and the electrical permittivity of free space and nothing else.

Free space just means a total vacuum: there is nothing there. The permeability of a substance is a measure of how it responds to a magnetic field. Likewise, the permittivity of a substance is a measure of how it responds

71

to an electric field. The permeabilities and permittivities of all substances are measured relative to the free space values.

When Maxwell substituted the values for the free space permeability and permittivity into his expression for the wave speed, he found a value which was equal to the measured speed of light. On the basis of this remarkable coincidence, he concluded that electromagnetic waves and light were one and the same. It was not until Heinrich Hess produced electromagnetic waves in his laboratory in 1887, that Maxwell was proved right, some eight years after his death.

Of course all this work was done long after the time of Galileo and Newton, for both of whom 'physics' was just the topic of 'mechanics'. However, in the eighteenth and nineteenth centuries, physics expanded to include thermodynamics, along with electricity and magnetism. But Maxwell's theory posed a problem for the older theories. According to Galilean relativity, Newton's laws took the same form in all inertial frames. This turned out (apparently) to be not the case for Maxwell's equations.

This is a crucial point and the word 'apparently' is important. In making a statement like this, we have to specify the way in which we test it. In Galilean relativity, we test it by making what are called Galilean transformations between inertial frames. Such a transformation is essentially the Galilean formula for the addition of velocities. When we make this transformation, Newton's laws of motion are of the same form in all inertial frames. (At the same time, we should qualify this statement. As we saw in the last chapter, two of Newton's laws are really general qualitative statements and hence are trivially true in general. These are N1 and N3. This leaves us with the nontrivial result that the second law N2 is invariant under a Galilean transformation. We call this property of N2 'Galilean invariance'.)

At this stage there is now a conflict between Newton and Maxwell and it needs to be resolved. Maxwell's equations are not Galilean invariant. It turns out that they are Lorentz invariant: that is if we replace the Galilean transformation between inertial frames by something called the Lorentz transformation, then Maxwell's equations are the same in all inertial frames.

At the most fundamental level, Newton is wrong and Maxwell is right. We shall come on to that in a later chapter, and explain the nature of the conflict and how Einstein resolved it. However, there is also a clue in Maxwell's result for the wave velocity of light that foreshadows Einstein's most extraordinary assumption which he made in setting up the special theory of relativity.

Chapter 10

Michelson-Morley on the travelator

As the famous Michelson-Morley experiment was a measurement the speed of light, we begin with a few remarks about the nature of light and, in particular, reason motive for doing the experiment.

Light is a particular form of electromagnetic wave or radiation and is part of a spectrum which extends from gamma rays to long-wave radio. We can illustrate this as follows. A nanometre or nm is a billionth part of a metre and for visible light the wavelength of the electromagnetic wave is a few hundred nm. Then the spectrum looks something like this:

1. Gamma rays: wavelength = one millionth part of 1nm.

2. X-rays: wavelength = about 1nm.

3. Visible light: wavelength = 400 - 700 nm.

4. Short-wave radio: wavelength about 1cm.

5. FM radio and TV: wavelength about 1m.

6. Medium- and long-wave radio: wavelength about 1km.

As all these waves travel at the same speed, their frequencies depend inversely on the wavelength, so that there is also a frequency spectrum, as well as a wavelength spectrum. Gamma rays have incredibly high frequencies, being about a million, million, million times higher than the relatively low-frequency radio waves. Typically, for long-wave radio, these are of the order of tens of kilohertz, so that they overlap the audible frequency spectrum for sound waves. Note that one kilohertz is one thousand cycles per second.

Visible light is transmitted through substances which we describe as 'transparent'. The speed of the light waves depends on the nature of the substance, with light being fastest in a vacuum, and next fastest in air. The speed of light in a substance, say glass or plastic, determines that substance's 'refractive index', which is the ratio of the speed of light in a vacuum to the speed of light in the substance. Typically, glass has a refractive index of about 1.3 - 1.8, depending on the type of glass. Those of us who need to wear strong spectacles can benefit from new plastics which have very high refractive indices, thus allowing us to have lenses which are relatively thin and light. So much better than the 'bottoms of glass bottles' of yore! While air under standard conditions has a refractive index of about 1.003, in practice air can be taken for 'vacuum', which is convenient for doing measurements on optical components such as lenses. However, one must bear in mind that, for fundamental physics, it is necessary to draw a distinction between the speed of light in a vacuum and that in air (or indeed any gas).

By the early 1800s, the wave theory of light was superseding Newton's 'stream of particles' theory. (But quantum physics brings us back to particles and the 'wave-particle' duality!) A major factor here was the work of Thomas Young, who pointed out that light could be a transverse wave rather than a longitudinal wave, hence bringing in the concept of 'polarization' which could explain optical effects such as birefringence (see the chapter endnotes). But the clinching detail was the theoretical work of Maxwell who, as we have seen, showed that light was a transverse electromagnetic wave. As waves needed a medium to wave in, like gases, liquids and solids for sounds; or water for water waves; this led to a revival of the ancient Greek idea of the 'aether' (the purer air above the atmosphere, that the gods breathed) which was reinvented as the 'luminiferous aether': a medium for the propagation of light. This was pretty weird stuff. It pervaded the universe, and hence must be so etherial that it did not impede the motion of the planets; yet was sufficiently 'rigid', that it could support transverse waves of high speed and frequency. (Whether the 'vacuum' or the 'space-time continuum' is any less weird is something that we shall not consider here.) Various attempts were made to detect the aether around the end of the nineteenth/beginning of the twentieth centuries, but nothing was found to support the concept and the idea has now been totally discarded.

However, one such experiment has achieved lasting fame as the 'most important null result in the history of physics'. This was the measurement of the speed of light by Michelson and Morley in (1887). They began by taking measurements in the direction of the Earth's orbital motion round the Sun; and then at right-angles to that direction. They obtained the same result for both directions and hence failed to detect the 'aether wind' which was expected to occur due to the Earth's motion through the hypothetical aether. Hence the conclusion: no aether wind, no aether.

Nevertheless, this was a crucial result. Because, if we discard the concept of the aether (which we do!), then the experiment tells us that the speed of light is the same in both directions, irrespective of whether the light source is moving or not. It is also a surprising result, indeed to physicists a shocking result. This is for two reasons.

First, because there is abundant evidence that for other waves, such as water waves or sound waves, the speed of the wave does depend on the speed of its source.

Secondly, the result is also clearly contrary to the Galilean relativity of velocities, which we discussed earlier, in the context of the travelator. In the case of the Michelson-Morley investigation, the first part of the experiment measures the speed of light in a moving frame.

That is, Earth is moving in the same direction as the light and hence is a moving frame. In the second part of the experiment, Earth is not moving in the direction of the light (recall: the light is directed at right angles to the Earth's orbit) and so is a fixed frame. Hence, according to Galilean relativity, there should be a difference between the two values measured for the speed of light. So let us, in imagination, put Michelson and Morley on the travelator, as it were.

In fact we will put just one of them on the travelator. The Michelson-Morley experiment can be realized as follows. Let us picture two identical corridors, of equal length and meeting at right angles. One of them is traversed by a travelator, the other isn't. In order to set up the experiment, Michelson at first ignores the travelator and walks along his corridor, as does Morley along his. They have trained themselves to walk at exactly the same steady pace and they verify this by setting out at the same time and arriving simultaneously at the intersection of the two corridors. Now they repeat the experiment with only one difference. Michelson steps onto the travelator and arrives at the intersection before Morley. We can work out their speeds by dividing the distance travelled by the time taken. Michelson arrived first, so had the higher speed. But that was just what we expected from Galilean addition of velocities.

Suppose both men walked at 4 mph relative to the ground, while the travelator moved at 5 mph relative to the ground. Then, during the experiment, Michelson moved at $4 + 5 = 9$ mph relative to the ground, so of course he arrived before Morley who moved at 4 mph relative to the ground. This result can be generalized to any speed of 'walking' and any speed of the travelator. In the actual Michelson-Morley experiment, the equivalent of a person walking was the light wave travelling, while the equivalent of the travelator was the motion of the Earth in its orbit round the sun.

In the next chapter we will consider the *ad hoc* attempts to explain the null result of the Michelson-Morley experiment. Although Einstein claimed to be unaware of this experiment, it is nevertheless helpful to make a *post hoc* connection with Einstein's theory of special relativity.

Chapter 11

The Fitzgerald contraction and the Lorentz transformations

The Michelson-Morley experiment was based on Michelson's interferometer which split a beam of light into two beams at right-angles to each other. Each beam was allowed to travel to a mirror which reflected it back, so that the reflected beams could be recombined. The basic idea was that if one beam had travelled faster than the other, then it would arrive back slightly out of phase and this would show up in the form of interference fringes. That is, it would show alternating light and dark rings when viewed on a screen. Alternatively, if one 'arm' of the interferometer were shorter than the other, then one could get the same result (i.e. interference) from this cause alone. Then if one had both effects present, they could cancel out and thus one would get the null result that was actually observed.

For this reason, the Irish physicist George Francis Fitzgerald proposed in 1889 that moving bodies were shortened in the direction of their motion by a factor which depended on their speed and the speed of light. On this basis it could be argued that the arm of the interferometer which pointed in the direction of the Earth's motion was shortened to such an extent that it just cancelled out the change in the speed of light. This effect was known as the Fitzgerald contraction, but is now more usually referred to as the Lorentz-Fitzgerald contraction, while the correction factor is often referred to as the gamma factor (after the Greek letter normally used to denote it) or the Lorentz factor. It is worth pointing out that this *ad hoc* modification of Galilean relativity was the first indication that the speed of light *in vacuo* is the largest speed that there can be.

Although Lorentz came after Fitzgerald, he made a more comprehensive (if still *ad hoc*) investigation of the null result of the Michelson-Morley experiment and in the process developed all the mathematical results that Einstein needed to set up the theory of Special Relativity.

Now we need to be a little more formal, and discuss the relationships between frames in relative motion in terms of transformations between them. The idea is easily illustrated by stating Galileo's transformations. So, sticking for the moment to our outdated terminology of 'moving frame' and 'fixed frame', let us recall that all inertial frames of reference are in uniform motion relative to some other frame. However, we can make the arbitrary choice of one frame as being 'fixed' (normally this is the frame that we are actually in), and we may state the relationship between two frames as follows:

1. The distance travelled (relative to the fixed frame) = The distance travelled in the moving frame + The distance travelled by the moving frame in the fixed frame.

2. Time goes at the same rate in all frames, fixed or moving.

It may be helpful to think of yourself as the 'moving body', the travelator as the moving frame and the airport terminal building as the fixed frame. Then it is obvious that these statements are true in terms of our everyday experience.

It is also the case that Newton's laws of motion are unaffected by these transformations and we therefore say that Newton's laws are Galilean invariant. However, if we apply this test to Maxwell's equations, they fail it. Maxwell's equations are not Galilean invariant.

The next step forward was due to the Dutchman, Hendrik Lorentz who, over the period 1892 to 1909, independently of Fitzgerald, came to the conclusion that the null result of Michelson-Morley required not only that the length of a moving body should be shortened in the direction of its motion, but that clocks on a moving body would also run more slowly. The outcome of his work can be stated in the form of two transformations, thus:

1. The distance travelled (relative to the fixed frame) = gamma x (The distance travelled in the moving frame + The distance travelled by the moving frame in the fixed frame).

2. An interval of time in the moving frame = gamma x (An interval of time in the fixed frame).

The factor gamma depends on the speed of the moving frame (expressed as a fraction of the speed of light) and varies continuously from unity (when the moving frame is at rest) up to infinity (when the moving frame is going at the speed of light *in vacuo*. As mentioned above, this was the first intimation that the speed of light represented an upper limit.

So, at everyday speeds (by which we mean speeds much less than that of light), the Lorentz factor is effectively unity. The Lorentz transformations reduce to the Galilean form and Newton's laws are all right. But at speeds close to the speed of light, where the gamma factor takes a large value, we must use Lorentz transformations. Under these circumstances, Maxwell's equations are invariant but Newton's laws are not. Hence Newton's laws need some sort of modification to make them fit with the Lorentz transformations.

Thus, by the beginning of the twentieth century, Lorentz had provided the basic tools. Then the great French mathematician, Henri Poincaré (1854 - 1912), had refined them and formally stated the problem. Now it just remained for someone to solve it.

Chapter 12

Enter Albert Einstein!

The story of Albert Einstein is well-known, as indeed are the myths and speculations still surrounding his life. There are many books and articles about him, and some of these are listed in the endnotes for this chapter.

The prosaic facts are that he was born in 1879 in Ulm (near Munich), where his father owned an electrotechnical business; was schooled there; and completed his education at the Polytechnical School in Zurich. By 1902 he was married and had become a patent examiner (Expert Class III) in the Swiss Patent Office in Berne. During his time as a patent examiner he evidently had plenty of time to work on his theoretical ideas, and in 1905 he published three papers on, respectively, the photoelectric effect (in which he introduced the light quantum), the theory of Brownian motion, and special relativity. In fact it was the first of these for which he was later to receive the Nobel Prize for physics in 1921. However, in 1906, he received his PhD from Zurich University (at the third attempt) and was promoted to Expert Class II at the Patent Office. In 1908, he secured an appointment at Berne University and thereafter held academic appointments at various universities. In 1915 he completed his general theory of relativity and, when the prediction of this theory that gravity would bend light was confirmed in 1919, he became world famous. He died in 1955, aged 76, in Princeton, where he had spent his life since 1939 at the (then newly created) Institute for Advanced Studies.

We have previously mentioned the problem posed by the derivation of Maxwell's equations. We will now state this more formally and then state Einstein's two axioms, which together make up Special Relativity.

The basic problem in classical physics (i.e. of a relativistic nature) was an inconsistency which can be stated as follows:

A According to Galileo's relativity, Newton's laws have the correct properties for relating events in one frame of reference to another moving at constant speed with respect to it. (This property is known as Galilean invariance.)

B According to Galileo's relativity, Maxwell's equations for for the propagation of electromagnetic waves have incorrect properties. Or to put it in the jargon, they are NOT Galilean Invariant.

So we have to choose between Galileo and Newton, on the one hand; and Maxwell, on the other. In doing so, we need to keep in mind the fact that Maxwell's equations describe the propagation of light. In attempting to understand what Einstein did, we also need to bear in mind that Newton's laws are effectively correct for everyday speeds which are much less than the speed of light. Einstein's assumption was that Maxwell's equations were more generally correct, although this may not exactly be apparent at first sight, when we state his two axioms, as follows:

Einstein's Axiom 1: The laws of physics are the same in all reference frames moving at a constant speed, irrespective of the value of that speed.

Einstein's Axiom 2: The velocity of light *in vacuo* is the same in all reference frames moving at constant speed, irrespective of the value of that speed.

At this point we should pause for consideration. We have just made some very abstract general statements. This is characteristic of physics. It is what we do. But in order to understand them, any physicist worth his or her salt needs to think of them in terms of some specific situations. That is what we will do next.

First, we introduce a corollary to Einstein's axioms. This is just a restatement of Axiom 1. It goes as follows:

> Einstein's Corollary: No physical experiment can be used to tell whether an inertial frame of reference is moving or at rest (with respect to any other frame).

This statement is a real tease. You may think you can find ways round it but ultimately you can't! We shall return to this later. But, in the meantime, we shall round off this section by just saying a few things about Einstein's two axioms.

Taking them in reverse order, Einstein's second axiom was foreshadowed by the work of Maxwell, as discussed in Chapter Nine. Indeed some readers may feel that the second axiom is rather obvious, in view of what we said about Maxwell's work. That would be fair comment. But one thing I have to make clear is that, although natural from some points of view, it is, for physicists, really very surprising from another. This is not something that we physicists have to just get over! From the point of view of addition of velocities Einstein's second axiom remains counter-intuitive, and leads directly to all the amazing consequences of Special Relativity such as length-contraction, time-dilation and the mass-energy equivalence.

Axiom 1 is, by comparison with Axiom 2, apparently quite innocuous. In fact it is much the same as Galileo or Newton would have said. But there is one subtle difference. To Galileo or Newton, the laws of physics would have been the laws of motion of bodies - what we nowadays call 'classical mechanics'. However, by Einstein's time, the laws of physics also included the laws of electromagnetic fields (There are, of course, also the laws of thermodynamics. But, from our present point of view, these can be treated as being on a par with classical mechanics.), in the shape of Maxwell's equations. This is why I have used the word *physics*. It emphasises the subtle but vital distinction between Galileo's relativity and Einstein's apparent restatement of it.

With Einstein, everything changed. His role was, like that of Newton before him, to provide a unifying framework which replaced a patchwork of ad hoc proposals and basic conflicts with a simple universal theory. The starting point was, of course, his two axioms, particularly the second (and shocking!) one: that the speed of light *in vacuo* was invariant.

Actually, from one point of view, the second axiom (shocking or not) was, with hindsight, really quite obvious. If we go back to Maxwell's theory, as discussed in Chapter Nine, we noted that the it predicted that the speed of all electromagnetic waves *in vacuo* (including light) depended only on the electrical permittivity and the magnetic permeability of free space. These two quantities are universal constants and hence the speed of light has to be a universal constant too.

Anyway, accepting the second axiom, the next step is to use some smart but very simple mathematics. We can write down general transformations between inertial frames in relative motion. (This is what mathematics is so good at. But it doesn't by itself solve the problem.) Then we insist that these transformations respect Einstein's second axiom. This insistence fixes the form of certain unknown quantities in the formulae for the transformations. The resulting outcome is that these new transformations are none other than the Lorentz transformations.

Of course, Einstein must have realized in advance that he was trying to derive the Lorentz transformations. So did he guess the answer and work backwards to obtain it? Probably ... but what of it? It is only in elementary school that that is regarded as cheating. In the real world, with formidable problems to be solved, we need all the help we can get!

Using Einstein's special relativity, a new expression can be derived for the transformation of speeds between frames in relative motion. This replaces the Galilean formula, which we derived with the help of the travelator, and now includes a version of the Lorentz gamma factor which depends on both the speed of the body and the speed of the moving frame. The important thing to note, is that when we add velocities in this new relative way, the speed of light is indeed independent of the frame. So everything checks out and Axiom 2 is satisfied. Of course, if you now apply this relativistic formula back to the travelator, then your walking speed and the travelator speed are so slow compared to the speed of light, that the formula just reduces back to Galileo's version, and once again, all is well.

Chapter 13

Four-dimensional space-time and Newton's laws

It used to be common to encounter the term 'the fourth dimension', and this usually referred to time. However, with the formulation of the geometry of a four-dimensional space-time by Hermann Minkowski in 1908, the fourth dimension actually became a matter of practical physics, and Einstein found a ready-made basis for his work in relativity.

In order to make this idea work, we have to multiply time by the speed of light in order to get a distance. This puts it more on the same footing as each of the three space coordinates, which all represent distances. So then we have to picture three space axes at right angles to each other and a fourth axis (involving time) at right angles to the first three. Not an easy thing to picture, let alone achieve. In fact, the way one gets round this is to make the fourth dimension purely imaginary.

Now, by 'imaginary' we do not mean in the everyday sense, but in the technical sense of mathematics. In mathematics, 'pure imaginary' is related to the concept of a root of a negative number. Just as $1 \times 1 = 1$, we also have $-1 \times -1 = 1$. Hence, in order to deal with roots of negative numbers, we introduce $i \times i = -1$, where i is an 'imaginary number'. Then, we can represent the root of any negative number by writing the root of the corresponding positive number and multiplying it by i. For example, the square root of -9 is $3i$.

This simple device then leads on the subject of complex numbers, one part of which is real and the other imaginary. These can be added, subtracted, multiplied and divided, once one learns the basic rules.

A good way to envisage this (at least for addition) is to think of a system based on apples and oranges. A complex number is then so many apples plus some oranges. Obviously we cannot add apples to oranges or *vice versa*. But clearly we can add complex numbers by adding all the constituent apples together and all the constituent oranges, ending up with a new complex number of so many apples and so many oranges. Unfortunately this analogy is limited as it doesn't tell us how to do multiplcation, for instance. For that we need actual complex numbers where an apple is just a number and an orange is a number multiplied by i, the square root of minus one.

The concept of complex numbers, in turn, further

leads on the concept of four-vectors, as opposed to the ordinary vectors of three-dimensional Euclidian space, which should now be called three-vectors, in order to distinguish them. Unfortunately we cannot go any further along this route, without introducing the dreaded mathematics. So, suffice it to say that relativity took on its most profound significance in this new four-dimensional space, which obeyed rules which were easily understood generalizations of ordinary three-dimensional geometry.

For example, one result of using this formalism is that Lorentz transformations can now be interpreted as rotations in four-dimensional space. And, although this may not seem very exciting, in fact it is one of the first steps in the beginning of a formalism which underpins the whole of modern theoretical physics. This becomes particularly the case when we consider electromagnetism. Not only do Maxwell's equations take their most succinct and elegant form in Minkowski four-space, but the duality between electric and magnetic fields is seen as a matter of geometry, in which the four-space rotations (dependent of the speed of a moving body) determine how much of the electromagnetic field is electric and how much magnetic.

To put it another way, we now find that Faraday's law of electromagnetic induction and Ampere's law, which relate electric and magnetic fields (respectively) to the variation with time of magnetic and electric fields (respectively), are seen to correspond to rotations of vectors in four-dimensional space-time. Thus, at this point, we are developing a unified view of moving bodies and electromagnetics in which everything is reduced to the geometry of a four-dimensional space. What is really good about this is that the introduction of this space also provided the basis of a unification which included Newton's laws of motion. So that is what we now consider.

First of all, as we have pointed out earlier, two of Newton's laws of motion are qualitative in nature and hence are like conservation laws (or, indeed, general principles): always true. The first law is just the principle of inertia: a body continues in a state of rest or uniform motion unless acted upon by a force. This is true for all situations, regardless of the speed of the body. So this goes over into Special Relativity. The same applies to action and reaction being equal and opposite. That is a symmetry principle, and is always true, irrespective of the speed of the bodies.

So we are left with Newton's second law, usually stated as: 'the force on a body is equal to its mass multiplied by its acceleration'. As we have previously noted, this law is Galilean invariant but not Lorentz invariant. As a result, strictly speaking it is incompatible with Maxwell's equations. In this sense it is wrong. But, and it is a very big 'but' indeed, it is absolutely fine in practice for a huge number of applications ranging from molecular motion to planetary motion, with many things in between from billiards through cricket, baseball and ballistics, to aerodynamics. So, you are not going to notice the 'wrongness' of N2 unless you are dealing with speeds near the speed of light. (In fact, once upon a time, engineers working on the design of particle accelerators had to be persuaded to make relativistic corrections, as Einstein's relativity was not part of their education. Nowadays even the ubiquitous 'sat-nav' requires relativistic corrections.) It is also the case, that Einstein inherited a Newtonian framework which could be adapted from three-space to four-space. To explain this in detail would take us far beyond the scope of this book but we shall now consider the general approach.

Before we do it, however, it is now time to say goodbye to the old-fashioned terminology of 'fixed frame' and 'moving frame'. All frames are moving and we are simply choosing one which it is convenient for us to designate as being at rest. Usually this is the frame in which we are at rest, so we introduce the idea of the 'co-moving frame'.

In its co-moving frame, a body is at rest. We then say that any quantity measured in the co-moving frame is the 'proper' quantity of that body. For instance, proper length, proper time, proper mass, and so on. Then, by generalising everything (length, time, mass ...) so that it transforms by Lorentz transformation in four-space, we find that we can adapt Newton's second law to a four-space, Lorentz-invariant form. In the process, we do surprising things like combining three-momentum with energy to make something called four-momentum; or adding the mass-energy equivalence ($E = mc^2$, of course!) into the statement of conservation of energy.

In Special Relativity, we end up with N2 in four-space form, but with some bizarre features such as the acceleration is no longer necessarily in the same direction as the applied force; and the classical force is supplemented by its rate of doing work. There is, unsurprisingly, also a gamma (or Lorentz) factor involved. However, when we restrict our attention to speeds that are small compared to the speed of light, then the four-space version of Newton's second law simply reduces back to the classical three-space version, as formulated originally by Isaac himself. That is, in the same way as the Lorentz transformation reduces back to the Galilean form.

Chapter 14

General Relativity

Many years ago, when I was watching a badminton match on television, I was intrigued by the occasional view taken with the camera more or less vertically above the court. The idea was to show how the players covered the court during rallies, and this was undoubtedly interesting. However, I found that I was more intrigued by the two-dimensional representation of the three-dimensional motion of the shuttlecock. At times this was quite fascinating, as it was not immediately obvious what was going on.

Let us now conduct a thought experiment in a similar vein, and imagine watching a snooker match through a television camera mounted vertically above the table. This already quite two-dimensional game would be rendered even more so, and would appear to consist of discs moving over a flat surface. Picture one such disc stationary in the field of view. Another disc comes into view and either hits the first disc or misses it. These are the only two outcomes possible. Either the moving disc hits the stationary one, and both are deflected. Or, it misses, and continues in a straight line. This is our two-dimensional projection of three-dimensional Newtonian dynamics as exemplified by the game of snooker.

Now suppose that we repeat the experiment, but this time there is a different outcome. The moving disc is deflected from its path as it approaches the fixed disc, does not strike it, but goes on in a straight line at an angle to its initial path. How do we account for this behaviour? There is no collision but there appears to be some kind of action at a distance.

One explanation might be that the two discs are magnetised but if we rule this out then we are left with the idea that it may be some local property of the snooker table. And if we move the fixed disc to different positions and it always occurs, then we could say that there is indeed some local property of the snooker table which follows the stationary disc (or, in three dimensions, the ball).

The behaviour we are observing (in our imaginations) can be explained in three-dimensions by having a rubber sheet stretched over the billiard table, the stationary ball is made of lead and the moving ball is made of plastic. (Or ivory or whatever billiard balls are made of: certainly it will be much lighter than lead.) The heavy stationary ball will cause a dip in the elastic surface and the light moving ball will be affected by this dip.

It is in this way that Einstein explained gravitational attraction. A massive object, such as a planet or a star, distorts the space-time continuum in its vicinity. In other words, it causes a 'dip' and this dip affects the motion of other bodies. In fact any body will cause a distortion which will extend indefinitely through the space-time continuum, although the magnitude of this distortion will depend on the amount of matter in the body. In three dimensions, this is seen as the inverse-square law.

This theory of gravitation is part of Einstein's General Relativity, and it is what made him famous. The theory predicts that light will be deflected by a gravitational field and it was just this effect that was found during Eddington's observation of the solar eclipse in 1919. It also makes other predictions, where Newton's theory is not quite accurate enough, and these predictions have been confirmed by subsequent observations. The beauty of the theory is that it explains how gravity acts at a distance, in terms of the distortion of the space-time continuum extending away from the massive body in question. This is often summed up in the phrase: 'matter tells space how to curve, and space tells matter how to move'.

However, this theory brings with it some strange consequences, such as gravitational time-dilation. Or, in the extreme case, a massive body can collapse into itself and cause a singularity in the space-time continuum. Nothing can escape from this singularity, not even light, and this leads to its popular name of 'black hole'. Needless to say, the clocks are all stopped forever inside a black hole.

We shall return to the topic of black holes in Chapter Seventeen, and discuss the way in which General Relativity predicts their existence. Here we now concentrate on some fundamental aspects of the theory. We begin by going back to our basic definition of relativity, as discussed in Chapter One.

As we noted earlier, for Galileo (and Newton) motion was a relativistic concept. The speed of a body could only be measured relative to some other body. This idea was taken to a deeper level by philosophers such as Bishop Berkeley (1685-1753) and Ernst Mach (1838-1916), who thought that in an otherwise empty universe, there is simply no way of telling whether or not a body is in motion. Under these circumstances, the idea is literally meaningless.

Mach took the view that it was the presence of the mass of the universe that validated the concept of motion, and that this mass therefore constituted the ultimate inertial frame. As the mass of the universe is mainly concentrated in the fixed stars, then the fixed stars may be regarded as constituting this 'ultimate inertial frame'. This idea, stated in various ways, is often referred to as 'Mach's principle'. It was one of the ideas that influenced Einstein in his formulation of General Relativity, but it is seen by many as vague (and even to have mystical aspects) and is the subject of debate. Indeed, it can be argued that General Relativity is not in fact compatible with Mach's principle! So we mention it here for historical completeness, and go on to consider some of the other ideas which helped towards General Relativity.

The first of these is the idea that the gravitational mass and the inertial mass of a body are the same. This had been an experimental result from the work of Galileo onwards. If a body falls under the effect of gravity, then the force of gravity acting on it is controlled by its 'gravitational mass'. This must be equal (by Newton's second law) to its 'inertial mass' multiplied by its acceleration. It is an experimental result that all bodies fall to Earth under gravity with the same acceleration, which is denoted by 'g'. Hence it is experimentally the case that the two masses are found to be equal, within experimental error. When this is expressed as the principle of equivalence, it takes the form:

gravitational mass = inertial mass,

without any qualification about experimental error.

In General Relativity, Einstein generalised this to the concept that gravity can be regarded as an inertial force. From a series of thought experiments (known as the *lift experiments*), he argued that a ray of light shone across a lift which was moving at constant speed would describe a straight line. But, if the lift were accelerating, then the light ray would follow a parabolic path, just like projectiles do under the effect of gravity. Given that we are accustomed to using light rays to define straight lines, this raises the possibility of a curved space.

Another way of introducing the possibility of a curved space is to consider how we would make measurements of length in a non-inertial frame. If we consider a rotating frame, and use a standard measuring rod to measure lengths, then we have to take into account the effects of Fitzgerald contraction on the measuring rod. It can be deduced that a triangle measured when the turntable is a rest, will have curved sides when it is rotating. Of course, it would have to be rotating very rapidly indeed, to produce linear velocities comparable to the speed of light. Also, in order to detect such effects, we would have to employ astronomical distances. For this reason, General Relativity is most applicable on a cosmological scale. We will return to this aspect in Chapter Seventeen.

Chapter 15

Time dilation and the 'twins paradox'

The 'twins paradox' has been on the go since just after Einstein published his paper on special relativity, and continues to generate controversy to this day. A distinguished writer on the subject has referred to the possibility of a 'hidden emotional content' for some people, because the paradox is quite easily resolved. With some experience of student discussions over the years, supplemented by a recent look on the internet, my impression is that some people focus on the idea of one twin being older than another (Of course one twin is always older than the other because they cannot be born simultaneously! But here we mean significantly older.) to the exclusion of everything else; and actually have either little idea, or perhaps even rather odd ideas, of what is meant by Special Relativity. For this reason alone, I think it is a good idea to pension off the twins and discuss a physical experiment involving clocks.

First let us clear one possible source of confusion out of the way. There is widespread confusion in physics (and in science generally) about the meaning of the word 'paradox'. *A paradox is an apparent contradiction.*

One often sees statements (in other areas as well as relativity) that 'of course there is no paradox' or it is just an 'apparent paradox'. These statements, if true, would rather suggest that those scholars who wrote careful papers pointing out that there is a paradox have been wasting their time! But, in fact, statements such as these merely reflect ignorance of the meaning of language, let alone science. If we can reconcile the opposing elements of a paradox, then the paradox is said to have been resolved. Once a paradox has been resolved, its status as paradox (rather than a contradiction) has been confirmed.

Having discarded the eponymous twins, we can focus now on what actually matters: the phenomenon of time dilation. This is often stated as 'moving clocks run slow'. But that may be an unfortunate choice of words, if it suggests that speed has an effect on the clock. In fact speed has an effect on the passage of time. The effect has been established experimentally beyond any doubt but it is still instructive to look at the paradox and see how it is resolved. Essentially we are going to carry out a thought experiment using two identical clocks, labelled A and B. We have to think very carefully about how this experiment is carried out and about how we justify the statements which we make.

The experiment can be described as follows:

Stage 1: We begin by comparing the two clocks, side by side on Earth. We establish that they record the passage of time at the same rate.

Stage 2: Clock B is put on a spaceship which travels away from Earth, attaining speeds that are large enough for relativistic effects to be easily measurable.

Stage 3: Clock B returns to Earth and is compared to clock A. It is found that B is slow. It was moving relative to A and time was dilated for it.

Now I want to emphasise that that is all there is to it! That is the result we would get if we actually carried out the experiment as just described. So where does the apparent contradiction arise?

It arises as follows. If we wish, we can work out theoretically by how much the moving clock B is slow (when compared to A) at the end of the experiment. This is a first-year physics exercise and is just a matter of using the Lorentz transformation of time. This depends (through the gamma factor) on the velocity V of the clock B relative to A. The answer obtained in this way is quite unambiguous and in agreement with experiment.

But, some devil's advocate can object that, in the reference frame of the spaceship, it can look as if the Earth is moving away with velocity $-V$ and so, according to Lorentz transformation (In the terminology of mathematical physics, the Lorentz transformations are symmetrical under interchange of V and $-V$.), clock A should run slow!

The crucial question here is this: which clock moved relative to the other, between the initial and final comparisons? The answer is clear: clock B on board the spaceship is the one which moved. It experienced an acceleration (which is NOT relative) as it left the Earth and a negative acceleration as it returned. The fact that it turned round or that its journey may have been in two parts has nothing to do with it. It is the fact that it experienced an acceleration, which everyone on the spaceship would have noticed, which distinguishes it from the clock which remained on Earth.

In physics we would say that the acceleration is 'symmetry breaking'. It breaks the apparent symmetry between the two frames, which only applies when they are both in uniform motion. It also destroys the apparent contradiction. The outcome of the experiment is as stated above and is fully in agreement with the theory.

So, what about the twins? Well, if we substitute 'twin' for 'clock', then twin B becomes younger than twin A by the end of the experiment. That is what appears to trouble people. It just seems to be so counter-intuitive. Some emotional factor stops people seeing that the whole experiment is very extreme in physical terms (it has never been carried out with living creatures), and the point at issue is time dilation, so there is no reason for its outcome to be intuitively obvious. Really, it is better to stick with clocks. Because a conundrum that was put up as a debating point for physicists at the beginning of the twentieth century has been troubling non-physicists ever since!

Chapter 16

Mass-energy equivalence and nuclear binding energy

At an early stage it was realised that Einstein's mass-energy equivalence provided an explanation (at least on energetic grounds) of how positively charged particles could overcome their mutual repulsion in order to congregate together in the nucleus; something which is referred to as 'nuclear binding'. This led on to the concept of obtaining energy from nuclear fission and nuclear fusion; the latter being the mechanism by which the Sun and other stars generate heat. In order to appreciate the nature of the problem, we begin with a brief account of the way in which the concept of the nuclear atom developed.

The nuclear age began in Manchester University in

1909, when the team led by Ernest Rutherford were shooting alpha particles into thin gold foils, and observed that some of the alpha particles actually bounced back, instead of passing through. Recalling it later, Rutherford said: 'It was almost as incredible as if you fired a fifteen-inch shell at a piece of tissue paper and it came back and hit you'. From this he deduced that most of the mass of an atom must be concentrated in a small volume at its centre. This was the nucleus (from the Latin *nux* meaning nut).

If the nuclear age began with Rutherford in 1909, then the atomic age began, either with Democritus in 400 BC, for giving us the concept; or with John Dalton, in the early nineteenth century, for showing us its relevance. Either way, we have ended up with the chemical elements which make up the periodic table, ranging from the lightest, hydrogen, with an atomic number of 1 to uranium as the heaviest, with an atomic number of 92.

In 1896 J. J. Thomson established that cathode rays (as produced by a heated metal filament in a vacuum tube) were in fact particles which were electrically charged. Later these particles were named electrons and it was established that they were about one thousandth of the mass of a hydrogen atom. Ultimately it was found that the beta particles, which were emitted by radioactive substances, were also electrons, thus making a connection between what came out of atoms naturally and what was being produced by electrical discharges in the laboratory. This led on to the first picture of what lay inside the atom: the so-called 'plum pudding' model which was put forward by Thomson.

Thomson started from the fact that atoms are electrically neutral. Hence, if atoms contain electrons, they must also contain some positive charge in order to cancel out the negative charge of the electrons. Also, the number of electrons in an atom would correspond to its atomic number. So he pictured an atom as a diffuse, positively charged material, which was studded with the small, light electrons, rather like currants in a pudding. If we take an electron to have one unit of charge, then hydrogen would be a blob of positive jelly with unit positive charge with a single electron in it; helium would be a heavier blob of jelly with two units of positive charge and two electrons in it; All the way up to uranium as a truly massive blob of jelly with 92 units of positive charge and 92 electrons scattered about in it.

This model was never going to explain everything. But its downfall was Rutherford's alpha particle scattering experiments. The concept of a nucleus led on to the planetary model of the atom, as proposed by Rutherford, who established that the nucleus of the hydrogen atom was a particle in its own right. This positively charged particle was named the 'proton'. So now we have a picture of the hydrogen atom as a massive, positively charged proton being orbited by a light, negatively charged electron. Drawing an analogy with planetary motion, the electron (planet) would be attracted by the positively charged proton (sun), and this electrostatic attraction (analogous to gravity) would be balanced (just as for the planets) by the centrifugal force.

The obvious generalisation of this model was to assume that the next element in the periodic table (Helium) has a nucleus of two protons, with two electrons orbiting around it. And so on, to higher atomic numbers. But in fact there are two serious problems with this picture.

First, classical physics would predict that an orbiting electron would lose energy by electromagnetic radiation, and would spiral into the nucleus. Secondly, two or more protons in a nucleus would fly apart because of electrostatic repulsion (i.e. like charges repel). So on two counts, this was a very unstable model of an atom. There was also a third, less obvious, problem concerning the atomic mass, and that was discovered by the chemists, rather than the physicists.

The first of these problems was solved by Neils Bohr, who arrived in Manchester, aged only twenty-five, in 1911. He was what nowadays we would call a 'postdoc'. In other words, still very much serving his apprenticeship as a theoretical physicist. Rutherford, who was a famously grumpy character took to him immediately, apparently describing him as 'the most intelligent chap I have ever met'.

Bohr saw the snag with Rutherford's picture of the electrons orbiting the nucleus at once, and solved it by making the radical postulate that electron orbits were quantised. That is, an electron could only travel in certain discrete orbits, in which they did not radiate or absorb energy. However, they could emit or absorb energy by jumping between the discrete allowed levels: the so-called quantum jumps. Later on, quantum mechanics was developed and explained why these discrete orbits existed. But, for the moment, that was the problem solved.

In making this proposal, Bohr was drawing on the quantum theory of Max Planck who, at the last Christmas meeting of the nineteenth century of the German Physical Society, put forward the radical idea that heat radiation was not a continuous wave, but rather consisted of discrete packages of energy. Each package contained a fixed quantum of energy which was determined by the frequency of the radiation and a constant of proportionality, now known as Planck's constant and always represented by the symbol h. We can write this as an equation in words:

> Quantum of energy = h x the frequency of the radiation.

Then the event that really transformed the status of the quantum came a mere five years later. This was in 1905, when Einstein used Planck's theory to explain the rather puzzling photoelectric effect. This effect takes place when light is shone on a substance and that substance becomes positively charged.

Even at this stage of its development, the Bohr's modification of the Rutherford model of the atom explained many of the experimental observations which had been obtained at that time by studying the light emitted by heated gases and vapours. So that too, added to the status of the quantum theory.

Bohr's theory solved the first of the three problems with the Rutherford model, which we discussed earlier.

111

The result was the improved Rutherford-Bohr model of the nuclear atom. However, attempts to extend this new model to any element which was more complicated than hydrogen encountered the two remaining problems. At the phenomenological level, the chemists were running into difficulties with their classification of elements by atomic weight. Unfortunately the Rutherford-Bohr model said that atomic number of an atom (being the number of protons in its nucleus) was simply related to its atomic weight (being the total weight of the protons in the nucleus). But experiments painted a more complicated picture.

First of all, there was a problem with Helium. If hydrogen has atomic number 1 and atomic weight 1, then helium should have atomic number 2 and atomic weight 2. Unfortunately, the experimental fact is that helium has an atomic weight of about 4. This was a distinct snag and led Rutherford to conjecture that there was another subatomic particle, with the same mass as the proton but without charge. He called this the *neutron*. Then the Rutherford nucleus, for any element, consisted of a number of protons (the atomic number), which had to be equal to the number of electrons orbiting round it, and a number of neutrons which was broadly similar to the number of protons but not necessarily equal to it.

Once we have introduced the neutron, then we can introduce the concept of 'isotopes'. The chemical nature of an element is determined by the number of protons in its nucleus but its atomic weight is determined by the number of neutrons. For example, hydrogen can have a neutron and the resulting form has an atomic weight of 2 and is known as heavy hydrogen or deuterium. It can also have a second neutron and is then known as tritium. These are the naturally occurring isotopes of hydrogen. It is possible to synthesise hydrogen with higher atomic weights by putting in more neutrons but such isotopes are highly unstable.

If we go to the other end of the periodic table, uranium with 92 protons has the largest naturally occurring atomic number and exists in forms uranium-234, uranium-235 and uranium-238, with number of neutrons 142, 143 and 146, respectively. Uranium-235 is fissile which means that it can undergo fission when bombarded with low-energy neutrons and is the form used in reactors and weapons. As it is only a small fraction of the naturally occurring element, it has to be separated out in order to obtain the necessary quantities.

The introduction of the neutron nicely cleared up the observed differences between atomic number and atomic weight. Now the next objective was to understand how the protons are persuaded to stick together, when their positive charges want them to fly apart. This is where Einstein's energy-mass equivalence comes in. The development of mass spectrometry in the late nineteenth century (especially by J. J. Thomson, who discovered the electron in this way) allowed atomic masses to be measured with great accuracy. Once these measurements had been done a most extraordinary result emerged. This is so important that we shall highlight it as follows:

> In every case, and for every element, the mea-
> sured mass of the nucleus is less than the
> sum of the masses of the nucleons which make
> it up.

So, in forming a nucleus from neutrons and protons, some mass has been lost. This is called the 'mass defect'. Every isotope of every element has its own mass defect. And by Einstein's famous law, $E = mc^2$, this mass defect corresponds to an energy defect. So by combining together in a nucleus, the constituent protons and neutrons are in a lower energy state than if they were separate. This energy defect is the binding energy which holds the nucleus together.

In physics we would say that the nucleons (i.e. the protons and neutrons) are confined to a potential well, just like a marble in a fruit bowl or indeed like the Moon going round the Earth. But in the nucleus the force is not gravity, which would be much too weak. It is a new force which is called the 'strong nuclear force'. So, with the existence of a cohesive nuclear force postulated, we simply say that a nucleus will be in its lowest energy state. This only tells us that it is possible: it doesn't tell us how it comes about. For the reader who is interested in learning more, there is a brief account, beginning with the 1932 theory of Yukawa, in the notes for this chapter at the end of the book.

Returning now to how we get energy from the nucleus, the general position is that any nuclear reaction which leads to a smaller mass defect in the final state when compared with the initial state will give out energy. Although the change in mass defect may be small, remember that the associated energy is given by $E = mc^2$ and that the speed of light squared is a very large number. As is the number of atoms in a few grams of any substance. So the energy release can be enormous and the question then is: how can one bring this about?

Nuclear fission was discovered in 1939 by Hahn and Strassman who bombarded uranium with neutrons. By studying cloud chamber photographs they found that the uranium nucleus broke up into two fragments (typically barium with atomic number 56 and krypton with atomic number 36) travelling away from each other with enormous speed. Note that the atomic numbers of the two fission fragments add up to 92, the atomic number of uranium. The essential point here is that the mass of the uranium nucleus is greater than the sum of the masses of the fission fragments and the lost mass appears in the equation as the kinetic energy of the fission fragments.

It was later discovered that there were some spare neutrons left over in each fission. So these could go on and hit more uranium nuclei, thus giving a chain reaction. In fact such chain reactions are the basis of both weapons and reactors. If a piece of uranium is large enough then there is a greater chance of a neutron striking another nucleus rather than escaping from the lump of metal. This is known as critical mass and a bomb is made by bringing two or more pieces of subcritical mass together, using a conventional explosive, to make a supercritical mass of uranium which then explodes.

A fission reactor, as used for generating electrical power, works rather differently. The uranium is spread out in small, subcritical masses and the space in between is filled by a moderator (heavy water or graphite) which slows the neutrons down and allows the nuclear fissions to proceed in a controlled way. Of course it is possible that a conventional explosion can occur, with some release of radioactivity, although there have been few such incidents. But, without wishing to minimise the seriousness of any such explosion, it is important to realize that a properly designed reactor can not turn into an atomic bomb. This also applies to fast reactors which do not use moderators. It is actually quite an impressive technical feat to make an atomic explosion!

For nuclear fusion to occur, we have to go to the opposite end of the periodic table and fuse hydrogen atoms together to make helium. It is actually necessary to have a complicated series of interactions known as the proton-proton chain to do this, but the essential point is that at the end of the process there should be less mass than there was at the beginning. This extra mass is given out as energy. The difficulty in making this happen is that we have to persuade protons to get close enough to overcome their electrostatic repulsion. This involves enormously high temperatures, such as are found naturally in the Sun and other stars, where this is how energy is generated. On Earth the only really successful fusion reactions have been in the hydrogen-bomb but work continues on the harnessing of this energy for electricity generation. It is vastly more difficult than harnessing fission power, but if successful should take care of our energy needs for a long time to come.

Chapter 17

Black holes

One of the few aspects of theoretical physics which has entered the popular consciousness, is the idea that Einstein's theory predicts the existence of 'black holes' in the space-time continuum. The notion of a region of space where even light rays vanish, never to be seen again, is undoubtedly intriguing. It seems to be very much in the spirit of the legend 'Here be dragons', which once decorated maps and indicated very clearly the limits of the cartographer's knowledge. Presumably science fiction has played a part in raising the popular awareness of black holes, but it has to be said that they are also regarded with some fascination even within the physics community.

In this chapter we will discuss the way in which Einstein's theory of General Relativity can lead to this prediction, and in order to do so, it will be found helpful to consider how Einstein differed from Newton. So this is where we will begin: with Isaac Newton and the apple!

It is one of the abiding myths of physics, along with Galileo supposedly dropping marbles from the leaning tower of Pisa, that Newton was inspired to discover the law of gravitation by the shock of an apple falling on his head. In fact the myth is not that far out, because Newton himself said that the sight of an apple falling had given him the idea. Recently, the Royal Society has made the biography of Newton by William Stukely available on-line. From this one finds that Stukely was with Newton when the apple fell, and describes the incident. Again, the interesting question is: what stood in the way of seeing the law of gravitation? This time it was not friction, as discussed for the laws of motion in Chapter Eight, but something much more subtle.

In essence, Newton was trying to explain the picture of the universe that had emerged from the telescopic observations of Galileo; and, in particular, Kepler's picture of the solar system. The problem was that he had to sort out gravity and the laws of motion, more or less simultaneously as a sort of interlocking puzzle. We can understand the particular flavour that gravity brought to this conundrum by invoking our present day knowledge and working backwards.

We all know that gravity is an attraction between masses. We may also know that it is described as an inverse square law. That is, the attractive force between two bodies depends on their masses and on the inverse square of the distance between them. So that the bigger the masses, the bigger the gravitational force between them. And the bigger the distance between them, the less the force, going as the square of the distance. For example, move two masses apart by a factor of two and the force between them lessens by a factor of four. And so on.

So let us see if we could verify this law by doing an experiment. If our two masses were mere points in space (and for a very large distance of separation this would be a good approximation for two spheres) then we can just measure the distance between the points. Or, what is the same, the distance between the centres of the spheres.

Now consider a very tiny sphere being attracted by a much bigger sphere. How do we take measurements? We could measure from the centre of the little sphere to the centre of the big sphere. But, what if the big sphere is of an appreciable size relative to the distance between centres? This would mean that the front of the big sphere was much closer than the distance between centres whereas the back of the big sphere was much further away. So which distance should we take as the distance between the two spheres?

Isaac Newton solved this problem by dividing the big sphere up into thin shells and working out the force between each shell and the small sphere. Then he added up the forces from all the thin shells, allowing the thickness of the shells to tend to zero, such that a limit was reached in which the set of shells became identical with the big sphere. That is, he invented a major branch of mathematics, known as 'integral calculus'. Not bad going for a man who had been deficient in mathematics as a student and who remedied his deficiencies by reading books which he bought at the Stourbridge Fair!

Now we come to the crux of the matter. In the case of an apple and the Earth, we are not talking 'small sphere, large sphere'. Instead, we are faced with a small sphere and an infinite plane surface. And under these circumstances, the force on the apple is constant. Objects fall to Earth with a constant acceleration. And, as we know from N2, a constant acceleration is equivalent to a constant force. So the inverse-square law itself is not immediately obvious by considering how things fall to Earth.

In fact, the inverse-square law was first proposed by the French astronomer Ismael Bullialdus, as an explanation of how the Sun 'seizes or holds the planets'. This was in 1645, when Newton would have been about three years old. Later the idea was suggested in a letter from Robert Hooke to Newton; and it was Newton who incorporated it into a mathematical theory of planetary motion. Later still, Hooke was to accuse Newton of plagiarism, a pattern which would be repeated.

If we are to consider the problem faced by all these scientists, then it will be helpful to throw our minds back to childhood, when we went fishing for minnows or sticklebacks, armed with a net and a jam jar on a string. Many of us will have experimented by swinging a jar of water round in circles and made the interesting discovery that the water did not fall out, even when the jar was upside down. At some stage we were told that this was an example of centrifugal force at work (some people regard centrifugal and coriolis forces as being fictitious forces: see the chapter endnotes for some remarks on this topic).

So, abandoning the jar of water as an unnecessary complication, a stone attached to a string would be a possible model for planetary motion. Just as the tension in the string balances the centrifugal force, so the gravitational pull of the Sun could be seen as balancing the centrifugal force acting on a planet.

The orbit of a stone swung round by a piece of string is a circle. And, as Kepler found, planetary orbits are ellipses. But, although the requisite maths is more complicated, this simple model can cope with ellipses as well. Essentially what happens, is that when the planet is further away from the Sun, the gravitational force is weaker, but the centrifugal force is also weaker, so the two forces remain in balance.

In this way, the inverse square law emerged from a consideration of the solar system, as did the equations of motion. Nowadays, for most physicists, this is what Newton's law of gravity means. For applications close to Earth, such as aircraft flight, ballistics or ballooning, the effect of the Earth's gravitational field on a body is a constant acceleration g. For applications further away from Earth, such as moons, artificial satellites or planetary dynamics, the inverse square law rules. This picture is completed by the Einstein field equations. Physicists are aware that they exist, but only theorists and, in particular, cosmologists would have any degree of familiarity with them. They are known to be frighteningly complicated but are needed for problems on a cosmological scale, which means distances of billions of light-years.

However, this is not the whole story of Newtonian

gravity. After Newton's time, the mathematicians Gauss and Poisson developed field equations from Newton's law. Later still, the discovery of the inverse square law for electric charges (i.e. Coulomb's law) led on the the development of field equations for electrostatics. This is a subject which is familiar to all physicists, in which the Poisson equation is used to calculate the variation of the electrostatic potential (i.e. the voltage, in everyday terms) for a given distribution of electric charges. Correspondingly, in Newtonian gravity, a Poisson equation is used to solve for the distribution of gravitational potential for a given distribution of matter.

For bodies like our Earth or Moon, the earlier Newtonian theory is perfectly adequate. But the Newtonian field equations are needed to calculate the gravitational field for gaseous planets or stars. Furthermore, it is possible to write down a Newtonian equation of motion to predict the behaviour of the cosmos. So the Newtonian field equations are of interest to astrophysicists and cosmologists. Where, then, does the difference between the Newtonian field equations and Einstein's field equations lie?

Recall that the gravitational Poisson equation allows one to solve for the gravitational potential in a three-dimensional space, given a particular distribution of mass. In contrast, the Einstein field equations operate in a four-dimensional space-time continuum, and solve (essentially) for curvature effects, given a particular distribution of mass-energy. When the Einstein equations are solved in certain cases the mathematics predicts what is called a singularity. This is what in physical terms is seen as a 'black hole'. The physical interpretation is one of gravitational collapse. Although gravity is a weak force, when a mass is large enough, the gravitational forces can become large enough to overcome all other forces. A large star which has cooled down can be modelled as a uniform, spherical cloud of dust. For this rather simple matter distribution, the Einstein field equations can be solved, with the prediction of a critical radius for gravitational collapse.

Without mathematics, this is as far as we can go. However, to conclude this chapter, it may be of interest to say something about the idea of a model universe which underpins any actual calculation or theory of the cosmos. A casual glance at the night sky will reveal a very complicated picture. How indeed does one model that? The answer lies in the so-called 'cosmological principle'. This an assumption that, on a sufficiently large scale, the universe looks the same in every direction (i.e. is isotropic) and over every distance (i.e. is homogeneous). By very large scale, what is meant is an average picture, in which detail is 'smeared out' by a partial averaging process. This partial averaging is over volumes with a diameter of billions of light-years. Although the scale is very different, this is much the same idea as averaging over the atomic or molecular structure substance in order to obtain a continuum picture of a gas, liquid or solid.

Chapter 18

The general unified formalism of physics

It is often said that quantum mechanics and relativity are the twin pillars of modern physics. By relativity, it is meant special relativity, although, as we have seen, relativity as such goes right back to Galileo and Newton. However, it is arguable that the great central pillar of modern physics is provided by classical mechanics, which is the formalism that developed from Newton's laws. In the eighteenth and nineteenth centuries, various mathematicians built on Newton's laws to develop new mathematical formalisms with the aim of achieving rigour and generality. Of these mathematicians, the two most significant were Joseph-Louis Lagrange (1736-1813) and William Rowan Hamilton (1805-1865).

If we lump their theories together, as the Hamilton-Lagrange formalism, then we can summarise it as a set of mathematical equations which can be applied to any mechanical system, and which reduces back to Newton's laws. However, unlike Newton's laws, the formalism relies on a general principle (Hamilton's principle of 'least action') and a formulation of the problem in terms of the total energy of the system. The strength of the general approach is that it is readily generalised to other topics such as electromagnetism or the theory of condensed matter (formerly, solid state and liquid physics). It is also readily generalised to the two extreme cases where the Newtonian description of nature breaks down. As we know, that is where speeds approach the speed of light; or where sizes and energies become so small that they are comparable to those of atoms (or smaller). We shall consider these two cases in turn.

If we consider speeds approaching the speed of light *in vacuo* then this is the territory of Special Relativity. We have already discussed the fact that Newton's laws can be modified to take account of this, but with the Hamilton-Lagrange formalism it is all so much smoother and more obvious. Hamilton's principle (because of its generality) will apply under all circumstances, and the total energy can be generalised by the use of Einstein's mass-energy equivalence. With a change to four-dimensional space-time and the introduction of the gamma factor (giving a property known as 'Lorentz invariance'), the relativistic version of the formulation can be derived. Then, just as the original formalism could be shown to lead to Newton's second law, the Lorentz-invariant formalism leads to Einstein's modified form of N2, but in a

126

less *ad hoc* way than as discussed earlier.

If we now go down to atomic and sub-atomic scales, then the formalism has to be modified by quantizing the total energy. But, at the same time, it has to be recognised that it no longer makes sense to specifiy instantaneous properties like position and momentum of individual particles. Instead we have to understand that we can only know about such things in an average sense.

To the extent that we need to consider individual behaviour in order to make theoretical predictions about a system, we have to replace ordinary numbers by what are called non-commuting operators. To explain this idea, let us consider two ordinary numbers which we call A and B. For a definite example, let us take $A = 2$ and $B = 4$. For ordinary numbers, it doesn't matter in which order you multiply them together. For instance, $AB = BA = 2 \times 4 = 4 \times 2 = 8$, in this simple example and we say that A and B commute. However, if we take $A =$ 'putting on your shirt', and $B =$ 'putting on your pullover', then clearly AB does not equal BA. (For those who know some calculus, A and B could be differential coefficients.) However, Hamilton's principle still applies, and the formalism which is equivalent to Newton's laws is readily modified to the atomic case. Indeed one can go further, and do the relativistic modification as well, so that we end up with relativistic quantum theory.

It is perhaps interesting to reflect that it is not actually Newton's laws, let alone Galileo's relativity, which breaks down when we go to the two extremes. It is our basic mathematical description of nature, which has served us so well for millennia (well, just about). That is to say, a description of the world in terms of our human scale and the measurements we make, be it setting out a building site, making an article of clothing or relating the motion of the planets to each other. In short, it is our basic Euclidian geometry which ultimately fails at very small scales and very high speeds. In the first case, space-time becomes discontinuous; in the second, space-time becomes distorted. But basic principles, such as conservation of energy remain the same. So rather than being pillars, I think that quantum mechanics and special relativity are buttresses to the main pillar of classical mechanics, which was built on Newton's laws.

We finish, as we began, with the question: what is special about Einstein's Special Relativity? Ignoring the tempting side-line, that there are those who regard the term itself as a misnomer (see the end notes for this chapter), we have seen that Einstein accepted Galileo's relativity in its entirety, with one exception. The speed of light *in vacuo* was to be regarded as universal and absolute. Everything is relative became everything except c is relative. Of course the restriction to the vacuum is essential. Light moves more slowly through any other medium and hence the speed of light in that medium is not an absolute quantity. This one assumption, which as we have seen, ran counter to the classical addition of velocities, led to a preference for Lorentz invariance (i.e. Maxwell's equations) over Galilean invariance (i.e. Newton's second law). Galilean transformations, although an excellent approximation in most everyday circumstances, were not strictly correct. It was in Maxwell's equations, which describe the propagation of electromagnetic waves, including light, that this showed up. At a stroke this also explained the null result of the Michelson-Morley experiment: light emitted on earth and travelling in the direction of the Earth's motion travels at absolutely the same speed as light travelling at right angles to it. That, as Einstein might have said, is relativity!

Chapter Endnotes and Bibliography

Chapter One

For a discussion of Galilean Relativity, see: 'Dynamics and Relativity' by W. D. McComb, Oxford University Press (1999).

As an indication of Einstein's significance as an icon, two rather bizarre examples come to mind. First, in Mel Brooks's comedy film 'High Anxiety', there is a scene where a small elderly man comes into the room. He looks like a caricature of Albert Einstein. The resulting joke depends on the fact that his Einstein-like appearance suggests that he is a professor. The character played by Brooks addresses him as 'Professor Little Old Man'. It turns out that he is actually Professor Lillolman!

My second example comes from The Oldie magazine, which has two crosswords, Genius and Moron. The icon for the Moron crossword is a sketch of a boy in a dunce's cap, whereas that for the Genius is a perfectly recognizable caricature of Einstein.

Chapter Three

Entropy is a concept which occurs in thermodynamics and was introduced to deal with the following conundrum. We know that energy is conserved, which means that it is never destroyed or changed in amount. However, from the studies of steam engines in the seventeenth century onwards, it was understood that machines ran down, despite the fact that the same amount of energy was present. Entropy was introduced as a measure of the unavailability of energy, meaning that although the energy was still there, it could no longer be used to perform useful work. In any irreversible process, the entropy can be shown to increase. So, increasing entropy, implies a loss of usefulness of energy. In time this was seen to apply to the universe as well. This fact is, in effect, the second law of thermodynamics.

Chapter Four

Some writers in antiquity also referred to the Sun and Moon as planets.

As regards Occam's Razor, I first encountered the term as the title of an exciting science fiction novel when I was still a student. Although, more than forty years on, I no longer remember anything about the story, I do vividly recall that it conveyed a sense of excitement about modern physics that if anything increased my enthusiasm for becoming a theorist. This early experience left me with a tendency to spell the name this way, although I later learned that it would be more correct to say 'Ockham's Razor', as it is associated with the name of the mediaeval logician and Franciscan friar, William of Ockham (1280 - 1349). Ockham, a tiny village in Surrey with a recorded history stretching back to the Domesday Book, is chiefly notable as being the birthplace of William of Ockham, although later the extraordinary Ada Lovelace lived at Ockham Park, in the early nineteenth century (see the end of this note).

William of Ockham was a formidable logician, known in his time as the 'More than Subtle Doctor', he ended up being excommunicated by Pope John XXII and lived out his days under the protection of Louis of Bavaria 'energetically promoting a separation of Church and State'. Evidently a very modern man! His eponymous 'razor' is usually quoted in Latin as *Entia non sunt multiplicanda praeter necessitatem*; while the English translation is given as: 'No more things should be presumed to exist than are absolutely necessary.'

It is also known as 'The Principle of Parsimony' and can be found in the work of other philosophers, going right back to Aristotle. Apparently, William did not refer to it as his razor, but it was so described by others, from the way in which he wielded it! There is a great deal of discussion in the literature about how Occam's razor relates to science, including the idea that it justifies ideas like the simplicity or beauty of scientific theories. However, remembering its origins in the interplay between logic and theology, it really just means that you shouldn't make things up in order to have a more convincing story! I suspect that most scientists obey it instinctively and without much conscious thought.

Although not so relevant to this book, Ada Lovelace was the daughter of the poet Lord Byron (famously described by Lady Caroline Lamb as 'Mad, bad and dangerous to know!') and was born in the year of the battles of Trafalgar and Waterloo. Her life was short and marred by illness, yet she is notable for being probably the first computer programmer and the first person to appreciate that computers could be used for more than just number-crunching. Her writings were based on the mechanical computer (or analytical engine) of Charles Babbage, which was never built. Like Babbage, she was several technological revolutions ahead or her time.

Chapter Five

Kepler's laws can be found in the book 'Dynamics and Relativity' by W. D. McComb (Oxford University Press, Oxford 1999), along with their derivation. They are stated as follows (K1 stands for Kepler's first law, and so on, and each statement is followed by a comment to help you interpret it.):

K1 The planets describe ellipses with the Sun as one focus.

Or: As the planets orbit round the sun they follow

an elliptical path. An ellipse is, unlike a circle, not characterized by its centre but rather by its focus.

K2 The radius vector drawn from the Sun to a planet sweeps out equal areas in equal times.

Or: As each planet moves round the Sun, an imaginary line drawn from the Sun to that planet sweeps out equal areas in equal times, irrespective of where the planet is in its orbital path.

K3 The squares of the periods of the different planets are proportional to the cubes of their respective mean distances from the Sun.

Or: The time taken for a planet to complete one orbit round the Sun (its period or its year) depends, in the way just stated, on the distance that planet is from the Sun. As the orbit is an ellipse, sometimes the planet is closer to the Sun and other times it is further away. So we have to take an average and use the mean distance.

Chapter Six

'Galileo's Daughter' by Dava Sobell (Fourth Estate, London 1999). At first I thought this an unpromising title; but, as I had enjoyed 'Longitude' by the same author, I decided to give it a try. In fact it is mainly about Galileo himself, and really brings his story to life. I enjoyed it very much.

Chapter Eight

'Isaac Newton: The Last Sorcerer' by Michael White (Fourth Estate, London 1997). Some years ago, I spent six months as a visiting fellow at the Isaac Newton Institute for Mathematical Sciences and, as I was visiting his institute, it seemed only civil to find out a bit more about the old chap. The paperback edition of this book had just come out, and it fulfilled the purpose admirably. There are many interesting anecdotes in it, but one which is rather touching is when Newton bought a glass triangular prism in 1664 at Stourbridge fair, in order to observe 'the celebrated 'Phenomenon of Colours'. So Newton did not actually discover this effect but merely wanted to see it for himself. When he got back to his rooms, he made a small hole in one of the window shutters in order to allow the sunlight to shine in. Then he placed the prism at this hole and observed the colours on the opposite wall. In his own words 'It was at first a very pleasing divertissement to view the vivid and intense colours produced thereby ...'

The Atomists were the adherents to the philosophies of the ancient Greeks Leucippius and Democritus (our word 'atom' comes from the Greek for 'uncuttable'). Our knowledge of them mainly comes from the poem by Lucretius, *De Rerum Natura*. or 'On the nature of things'. Their extremely rational view of the natural world languished until experimental evidence emerged for the atomic nature of matter. The first such evidence is usually taken to be the discovery of the motion of pollen grains in a liquid suspension by the botanist Robert Brown in 1827. The discovery of Brownian motion required a microscope, rather as the discovery of the nature of the planets required a telescope. One could hardly overestimate the importance of these two optical instruments to the development of science.

In 1983 the US space-craft Pioneer 10 became the first man-made object to leave the solar system.

Chapter Ten

Birefringence is double refraction and was originally found in naturally occurring crystals such as calcite and Iceland spar. Essentially it means that the refractive index can depend on the polarization or even direction of propagation of the light rays. A classic demonstration of the effect is a crystal sitting on some print showing the words iceland spar which are double-imaged through the crystal.

Chapter Twelve

This selection of books about Einstein just reflects what I happen to have on my bookshelves, but I suppose that is probably as good a criterion as any.

1. 'Albert Einstein' by Banesh Hoffmann, in collaboration with Helen Dukas (Hart-Davis, MacGibbon 1973).

2. 'The Genius of Science: A portrait gallery of the twentieth-century' by Abraham Pais (Oxfore University Press 2000). Pais is better known as the author of 'Subtle is the Lord. The science and life of Albert Einstein' but the present book also covers quite a few other scientists as well.

3. '$E = mc^2$: A Biography of the World's Most Famous Equation' by David Bodanis (Pan Books 2001). Explaining this equation is the theme which links together various topics which make up a general account of modern physics.

4. 'Variety of Men: Statesmen, scientists and writers' by C. P. Snow (Penguin Books 1969). The common factor here is that Snow had actually met all his subjects. As with all Snow's writing, it is eminently readable, but Einstein finds himself in some strange company here!

5. 'Who got Einstein's Office? Eccentricity and genius at the Princeton Institute for Advanced Study' by Ed Regis (Penguin Books 1989). This tells the story of the Institute rather than that of Einstein, but still sheds some light on the great man.

Chapter Sixteen

In fact the problem with nuclear binding at that time, was rather like the situation regarding cohesive forces between atoms in earlier times. However, nowadays the interactions between atoms can be understood in terms of quantum theory. So what about nuclear binding? The first step on the way to doing this was taken by Hidekei Yukawa in 1932, when he proposed that nucleons were held together by the exchange of a new subatomic particle between neutrons and protons.

In quantum theory, such a process would lead to a so-called 'exchange force'. A classical way of envisaging this is to imagine two dogs quarrelling over a bone. One snatches it, the other snatches it back, and so on. If they do this very rapidly, then they are locked together by the process. In the atom, the 'bone' is a new particle which Yukawa estimated would be about 200 times heavier than an electron and 10 times lighter than a proton.

This assumption did not inspire much enthusiasm among physicists until later in the same year Carl Anderson discovered that particles of this mass were contained in the cosmic rays which arrived on Earth from outer space. Later these particles were called 'mesons', and in time it was found that there were many different mesons, and in turn many other different subatomic particles. Nowadays the situation regarding such particles is not unlike that regarding the elements and the periodic table in the nineteenth century. That problem was ultimately solved by quantum theory. Currently attempts to explain the 'periodic table' of the subatomic particles in a comparable way come under the heading of 'particle theory'.

In the process of studying cosmic rays it was found that the half-life of the particles was too short for them to be arriving on Earth. This turned out to be the first verification of the time-dilation predicted by Einstein's Special Relativity. The mesons were travelling so fast that their decay process was slowed down.

Chapter Seventeen

Everyone is familiar with the idea of centrifugal force, as the force that we experience when cornering rapidly in a motor car or when on a fairground ride. The concept of coriolis force is much less familiar and I will explain it in a moment. What concerns me here is the confusion which has arisen about these concepts, in the way in which they are taught to scientists and engineers. The argument goes that, when we refer the motion of a body to a rotating reference frame (such as the Earth), then the mathematical formulation includes new terms which are described as representing apparent or fictitious forces. As a result, some teachers regard centrifugal force in particular as being fictitious.

I had this brought home to me many years ago when I was teaching a physics service course to engineers, chemists and life scientists. The engineers told me they couldn't do the tutorial examples because Mr So-and-so in Engineering didn't use centrifugal force. So I told them to use Mr So-and-so's methods to solve the problems. But apparently they couldn't do it that way either. In the end we agreed to adopt as 'course convention' that, when doing the work of my course, they would accept the concept of centrifugal force. So peace and harmony were restored!

Shorn of mathematics, the argument goes like this. A moving body wants to obey N1 and keep moving in a straight line at constant speed. We apply a force to it (e.g. tension in a string, gravitational attraction) and, as a result, the body is forced to move in a circle; and is therefore constantly accelerating towards the centre. This is known as 'centripetal' acceleration and is only an apparent acceleration because the moving body never actually goes in the direction of the centre. The acceleration is purely due to its changing its direction.

However, suppose we consider a laboratory centrifuge which whirls a test tube round in a circle at high speed, so that we can separate out (for instance) some solid particles suspended in a liquid. We begin with particles uniformly distributed over the fluid and end up with a sediment at the bottom of the tube and the clear liquid above it. Spin driers work in just the same way. As do salad spinners.

During the operation of the centrifuge, the test tube experiences this apparent (centripetal) acceleration towards the centre, whereas the solid particles in it experience a net force outwards and end up plastered against the bottom of the tube. So far as those particles are concerned, centrifugal force is very real.

Now, for Coriolis force. For this we need the sort of roundabout you find in a children's playground. If two children throw a ball back and forward over the roundabout, it will travel in a straight line, as viewed from above or directly below. Now consider what happens if the roundabout is rotating. The movement of the ball is totally unaffected. But a third child sitting on the roundabout would see the ball move in a curving path and would conclude (if intelligent) that there must be some force acting on the ball. That would definitely be a fictitious force. This fictitious force is known as Coriolis force.

But, is Coriolis force always a fictitious force? If the children change places, so that the two who were throwing the ball now stand on the roundabout and roll it to each other, while the third child stands on the playground, what does each observe? Well, surprisingly, each will observe what they observed before! But, as they have all changed reference frames. (This matters here because one of the two frames - the rotating roundabout - is non-inertial.) what they observe now has a different significance. The two children on the moving roundabout will see the ball move in a straight line, back and forward between them. Again the third child sees the ball move in a curved path. But this time relative to the playground rather than relative to the roundabout. So the ball is actually experiencing a real force. Due to friction it is being dragged round by the roundabout and in this case Coriolis force is surely a real force.

When I wrote my book 'Dynamics and Relativity', I took the view that centrifugal force should be regarded as real force and Coriolis force as fictitious. However, a friend pointed out that where railways run North-South, there is Coriolis force on the moving trains, due to the East-West rotation of the Earth. The result is extra wear on the insides of the rails. I don't have a reference for this nor for the comparable idea that the meanders of rivers which flow North-South also show an effect of Coriolis force. But clearly there are cases where it may be considered a real force, so it doesn't do to be dogmatic in these matters.

Chapter Eighteen

An article on Relativity in 'The Oxford Companion to the History of Science' (Ed. J. L. Heilbron, Oxford University Press: 2003) describes the special theory of relativity (*sic*) as having been 'misnamed' and later states that the 'infelicity of the term' refers to both words: 'special' and 'relativity'. The author feels that it should be 'limited' rather than 'special'; and what distinguishes it from classical mechanics is not 'relativity' but 'the startling concept' that all observers (irrespective of their relative velocities) measure the same speed of light in free space.

This view appears to be based on incomplete knowledge or understanding. That is, insofar that it is more than a matter of semantics (and hence personal opinion)! The general statement of Galilean Relativity is that the laws of *mechanics* are the same in all frames. Whereas the general statement of Special Relativity is that the laws of *physics* are the same in all frames. This distinction is not trivial. As we have noted earlier, Maxwell's equations are not Galilean invariant. They are Lorentz invariant and hence properly accommodated in Einstein's relativity but not in that of Galileo.

Turning to the matter of semantics, I wondered what Albert Einstein had actually said. So I asked a native German speaker for help. The original German is *Spezielle Relativitatstheorie*. Apparently this can be translated as either 'Special theory of relativity'; or 'Theory of special relativity'. Irrespective of the exact order, evidently Einstein meant both 'special' and 'relativity'. (I thank my student Moritz Linkmann for this information.) And this is how the theory is universally known in physics.